Assessment and Intervention with Children and Adolescents Who Misuse Fire

PROGRAMS THAT WORK

Assessment and Intervention with Children and Adolescents Who Misuse Fire

PRACTITIONER GUIDE

DAVID J. KOLKO
ERIC M. VERNBERG

OXFORD
UNIVERSITY PRESS

OXFORD
UNIVERSITY PRESS

Oxford University Press is a department of the University of Oxford. It furthers the University's objective of excellence in research, scholarship, and education by publishing worldwide. Oxford is a registered trade mark of Oxford University Press in the UK and certain other countries.

Published in the United States of America by Oxford University Press
198 Madison Avenue, New York, NY 10016, United States of America.

Library of Congress Cataloging-in-Publication Data
Names: Kolko, David J., author. | Vernberg, Eric M., author.
Title: Assessment and intervention with children and adolescents who misuse fire :
practitioner guide / David J. Kolko, Eric M. Vernberg.
Description: New York : Oxford University Press, 2017. |
Includes bibliographical references. Identifiers: LCCN 2017002681 (print) |
LCCN 2017005339 (ebook) | ISBN 9780190261191 (pbk.) |
ISBN 9780190261207 (updf) | ISBN 9780190670153 (epub)
Subjects: | MESH: Firesetting Behavior—diagnosis | Firesetting Behavior—therapy | Child | Adolescent
Classification: LCC RJ506.P95 (print) | LCC RJ506.P95 (ebook) |
NLM WM 190.5.I6 | DDC 616.85/843—dc23
LC record available at https://lccn.loc.gov/2017002681

In memory of my father, Myron Kolko
—David J. Kolko

About *Programs That Work*

Stunning developments in healthcare have taken place over the last several years, but many of our widely accepted interventions and strategies in mental health and behavioral medicine have been brought into question by research evidence as not only lacking benefit, but perhaps, inducing harm (Barlow, 2010). Other strategies have been proven effective using the best current standards of evidence, resulting in broad-based recommendations to make these practices more available to the public (McHugh & Barlow, 2012). Several recent developments are behind this revolution. First, we have arrived at a much deeper understanding of pathology, both psychological and physical, which has led to the development of new, more precisely targeted interventions. Second, our research methodologies have improved substantially, such that we have reduced threats to internal and external validity, making the outcomes more directly applicable to clinical situations. Third, governments around the world and healthcare systems and policymakers have decided that the quality of care should improve, that it should be evidence based, and that it is in the public's interest to ensure that this happens (Barlow, 2004; Institute of Medicine, 2001, 2015; Weisz & Kazdin, 2017).

Of course, the major stumbling block for clinicians everywhere is the accessibility of newly developed evidence-based psychological interventions. Workshops and books can go only so far in acquainting responsible and conscientious practitioners with the latest behavioral healthcare practices and their applicability to individual patients. This new series, Programs*ThatWork*™, is devoted to communicating these exciting new interventions to clinicians on the frontlines of practice.

The manuals and workbooks in this series contain step-by-step detailed procedures for assessing and treating specific problems and diagnoses. But this series also goes beyond the books and manuals by providing ancillary materials that will approximate the supervisory process in assisting practitioners in the implementation of these procedures in their practice.

In our emerging healthcare system, the growing consensus is that evidence-based practice offers the most responsible course of action for the mental health professional. All behavioral healthcare clinicians deeply desire to provide the best possible care for their patients. In this series, our aim is to close the dissemination and information gap and make that possible.

This Practitioner Guide presents a modular behavior intervention for working with children and adolescents who play with or set fires. Firesetting is a surprisingly common problem that can have devastating consequences for families and communities, but there has been limited practical, evidence-based information on how to address this behavior.

This comprehensive intervention is appropriate for school-aged children and adolescents and closely involves their parent(s) or caregiver(s). Treatment encourages more appropriate and safe behaviors, and caregivers are guided in limiting children's exposure to flammable materials. The treatment program is flexible so that the practitioner can decide which topics to cover to best meet the child's needs. This book includes an extensive appendix with separate forms and worksheets for the facilitator, parent/caregiver, and child.

Anne Marie Albano, Editor-in-Chief
David H. Barlow, Editor-in-Chief
Programs *ThatWork*

References

Barlow, D. H. (2004). Psychological treatments. *American Psychologist, 59,* 869–878.

Barlow, D. H. (2010). Negative effects from psychological treatments: A perspective. *American Psychologist, 65(2),* 13–20.

Institute of Medicine. (2001). *Crossing the quality chasm: A new health system for the 21st century.* Washington, DC: National Academy Press.

Institute of Medicine. (2015). *Psychosocial interventions for mental and substance use disorders: a framework for establishing evidence-based standards.* Washington, DC: National Academy Press

McHugh, R.K., & Barlow, D.H. (2012). *Dissemination and implementation of evidence-based psychological interventions.* Oxford: Oxford University Press.

Weisz, J. R., & Kazdin, A. E. (2017). *Evidence-based psychotherapies for children and adolescents* (3rd ed.). New York: Guilford.

Accessing Programs *ThatWork* Forms and Worksheets Online

Forms and worksheets from books in the PTW Series are made available digitally shortly following print publication. You may download, print, save, and digitally complete them as PDFs. To access the forms and worksheets, please visit http://www.oup.com/us/ttw.

Contents

Acknowledgments

For David Kolko, this book represents a long-standing commitment to the dissemination of tested practices and recommendations based on a professional career working with children who set fires that began nearly 40 years ago. This work began when I, somewhat naively, accepted the referral of a case of a young child who had set fires in his home and front yard after he was denied services by several local community agencies. At that time, there was virtually no empirical literature on firesetting in children or adolescents other than a handful of papers and reports describing the clinical features of some unique samples referred for evaluation. And, there were no evidence-supported clinical guidelines for managing firesetting behavior. Clearly, the absence of assessment and intervention guidelines created a significant gap that needed to be addressed. Consequently, grants, contracts, and service collaborations with supportive local and national colleagues were developed to enable us to conduct and publish scientific studies on various topics related to the description, assessment, follow-up, and treatment of this population in an effort to advance the knowledge base in this field. It is our sincere hope that this early work and this guidebook will encourage more practitioners to step up and serve this important population with materials and methods that we have found to work.

I am grateful to my co-author, Eric Morrison Vernberg, for inviting my collaboration on the implementation field trial where much of this material was applied and revised by many dedicated professionals. Likewise, I wish to acknowledge the collegiality, clinical contributions, and editorial assistance of the staff of the SAFETY Program at Western Psychiatric Institute and Clinic, especially Eunice Torres, Hillary Cohen, Jessica Gahr, and Kevin Rumbarger, and our dedicated students Samantha Walker, Jordan Harris, and Sam Fernbaugh. I owe a debt of gratitude to my early mentor, Dr. Alan E. Kazdin, and colleagues Drs. Brian Day and Claudia Roth, for their academic and clinical encouragement at the University of Pittsburgh, School of Medicine. Several other professionals in this country (Drs. Ken Fineman, Vaughn Hardesty, and Tim Kopet, Judy Okulitch,

Dr. Elizabeth Osborn, Irene Pinsonneault, Dr. George Sakheim, and Dr. David Wilcox) and in the United Kingdom (Joanna Foster, Robin Morris Jones, Tom Aitken, and Wendy Cooper) also deserve mention for their contributions to the field more generally. Our work would not have been possible without the support of grants from the National Institute of Mental Health (MH 39976) and the Allstate Foundation, and the assistance of the Pittsburgh Bureau of Fire. Our helpful editors at Oxford Press (Andrea Zekus and Sarah Harrington) and the Total Book (Kate Scheinman) made it possible for this book to reach the light of day. Finally, I am grateful to my wife, Judith, for supporting what I do, and to my children, Rachel and Aaron, for doing things I support.

For Eric Vernberg, the implementation project I directed that was carried out in the Kansas City metropolitan area between 2009 and 2014 contributed greatly to the refinement of the assessment and treatment approach described in this book. Many former graduate students from the Clinical Child Psychology Program at the University of Kansas helped by participating in data collection efforts, research team meetings, training and consultation activities, and the ongoing process of editing (and reediting) the manual throughout the implementation project. Ed M. Morris, Leni Swails, and Sonia Rubens deserve special recognition for helping with the initial and middle phases of the implementation project. Michelle Hendrickson and Bridget Cho contributed to the final phase of the implementation project. We thank the many administrators, clinicians, and staff at The Family Conservancy (headquartered in Kansas City, Kansas) for their interest and effort in the implementation project over those 5 years. Their experiences in using the assessment and treatment materials with fire-involved children and adolescents helped us in many ways as we refined this manual. We also thank the Kansas City (Missouri) Fire Department for supporting the implementation project throughout the entire process. Dr. Richard Gist helped immensely in organizing and supporting the ongoing collaboration between the fire departments in the Kansas City metropolitan area, the Family Conservancy, and the Clinical Child Psychology Program at the University of Kansas. Dr. Gist was instrumental in garnering financial support for the implementation trial through Fire Prevention and Safety Program grants from FEMA/US Dept of Homeland Security. These included: Award # EMW-20080FP-02405, 08-14-09 to 12-31-10; Award # EMW-2010-FP-01734, 06-20-11 to 6-19-12; and Award # EMW-2011-FP-01727, 08-31-12 to 8-30-13.

Assessment and Intervention with Children and Adolescents Who Misuse Fire

The Significance of the Problem

SCOPE AND SEVERITY

Firesetting by children or youth is a common and costly problem that has been found responsible for significant individual and societal consequences. By firesetting, we mean the unsanctioned use of fire that results in the burning of material, property, or a person. In some reports, children playing with fire may include actual firesetting behavior or related behaviors, such as playing with matches or lighters, which may not result in any consumption of materials or damages. Nearly 15 years ago, Hardesty and Gayton (2002) reported several statistics based on data from the National Fire Incident Reporting System to illustrate the breadth of the problem of children misusing fire. In 1997, children playing with fire started 65,000 fires that caused 284 civilian deaths and 2,158 civilian injuries. In 1993–1997, 86.3% of fatal victims of child-set fires were children between 0 and 17 years of age. In terms of FBI statistics, juvenile fire setters accounted for 52% of all arson arrests in 1998 and 45% of all arson offenses solved by arrest. In addition, 34.6% of all persons arrested for arson in 1998 were under 15 years of age, and 5.7% were under 10 years of age. Finally, arson had the highest rate of juvenile involvement in the FBI crime index. Such findings indicate that many child injuries and deaths are directly related to children playing with fire.

More recent data from the National Fire Protection Association (NFPA) for the period from 2005 to 2009 indicate that children were responsible for an average of 56,300 fires causing 110 civilian deaths, 880 civilian injuries, and $286 million in direct property damage per year (National Fire Protection Association, 2011). FBI statistics reported that nearly half (46%) of arson arrests between 2005 and 2006 were juveniles and nearly half of these were under 16 years of age. A total of 84% of the fires coded as "children playing" were set by children under the age of 10.

The numbers show continued improvement for the latest time period reported by the NFPA, namely, 2007–2011 (Campbell, 2014). Children set an average of 49,300 fires, caused 80 deaths and 860 injuries, and caused property damages estimated at $235 million per year. Structure fires were common (23% of all fires) and resulted in the most damages. In terms of the methods used, most were lighters (52%), followed by matches (18%), and candles (5%). Most fires were set in the bedroom (39%). Children less than 6 years of age often set home fires (43% of all home fires), whereas older children (10–12) more often set outside fires (39% of outside fires). In terms of arson arrests (Federal Bureau of Investigation, 2013), 27.9% of all arrests were for arson; this rate remains the highest juvenile rate for any crime. These data suggest that the incidence of juvenile firesetting has declined in the past decade; however, firesetting and arson among children and youth remain serious national health and social problems.

A few surveys of children or adolescents conducted over the past three decades provide an indication of the prevalence of fireplay and firesetting (Cole et al., 1983; Cotterall, McPhee, & Plecas, 1999; Kolko & Kazdin, 1988b; Terjestam & Ryden, 1996). School surveys generally report modest to high rates of self-reported firesetting during the past year for various cities including Rochester, NY (14% past year; 38% lifetime rate), Surrey, British Columbia (51%–62% for males and females), and Lund, Sweden (51%–62% for males and females). A survey of school children in community and clinic settings in Pittsburgh, PA (6–13 yrs) reported firesetting rates of 19% for outpatients and 35% for inpatients.

A recent report by the Illinois Department of Children and Family Services described a uniform assessment process at entry into state custody using a trauma-informed version of the Child and Adolescent Needs and Strengths tool (Lyons, McClelland, & Jordan, 2010). Based on information from these assessments, the authors reported an estimated prevalence rate of firesetting of 1.35% among 4,155 children and youth entering state custody. The rate for children and youth ages 10 years and older was 3.5%. The rate was higher for boys than girls but there was no relationship to racial background.

Characteristics of Children's Firesetting Incidents

We have learned some details of children's firesetting incidents based on studies that use structured interviews. Parents completed the Fire Incident Analysis to document parameters of their children's most serious incidents (Kolko & Kazdin, 1991b). Firesetters were classified as high and low on each of two primary motives (Curiosity, Anger). Heightened curiosity was associated with greater psychopathology, past fire involvement, and fire involvement 2 years later, whereas heightened anger was associated with behavior problems and later matchplay. Children from the same study completed the Fire Incident Analysis for Children (Kolko & Kazdin, 1994). The children acknowledged having access to incendiaries, limited remorse, limited parental consequences, being curious, and having fun. Four fire characteristics predicted their overall severity of involvement in fire at follow-up (i.e., fire was out of home, acknowledgement of being likely to set another fire, a neutral/positive reaction to the fire, no parental response to the fire).

Individual and Family Characteristics and Correlates

Standardized instruments have been developed to operationalize several potential fire-specific risk factors for firesetting, one for parents (the Firesetting Risk Inventory or FRI; Kolko & Kazdin, 1989a) and one for children (Children's Firesetting Interview or CFI; Kolko & Kazdin, 1989b). On the FRI, parents of firesetters (vs. those of nonfiresetters) report more significant concerns on several fire-specific factors (e.g., curiosity about fire, involvement in fire-related acts, exposure to peers/family fire models) and general child/parent (e.g., negative behavior) or family environment factors (e.g., use of harsh punishment, less effective mild punishment). On the CFI, firesetters (vs. nonfiresetters) have acknowledged more attraction to fire, past fireplay, family interest in fire, exposure to friends or family who smoke, and, somewhat surprisingly, knowledge of things that burn, but tended to show less fire competence (skill) on role-plays or interview questions.

Some clinical studies have also shown a relationship between childhood firesetting and child dysfunction, such as heightened aggression, psychopathology, and social skills deficits (Kolko, Kazdin, & Meyer, 1985). In a related study (Kolko & Kazdin, 1991a), firesetters were reported to exhibit more covert antisocial behavior than both matchplayers and nonfiresetters, and firesetters and matchplayers received more extreme scores than nonfiresetters on measures of aggression, externalizing behaviors, impulsivity, emotionality, and hostility, though they did not differ from one another. Child report measures found only a few differences associated with firesetters (e.g., aggression, unassertion, low self-esteem), relative to nonfiresetters. Some recent evidence suggests that child firesetters have many of the same characteristics as adult firesetters (Gannon & Pina, 2010).

A 2009 study by MacKay, Paglia-Boak, Henderson, Marton & Adlaf surveyed a total of 3,965 students from urban and rural communities across Ontario, ranging in age from 11 to 19 years (grades 7 to 12), of whom 52% were male. The youth were classified into four firesetting levels based on age of onset of their fire involvement and frequency of their current firesetting. Fire involvement was more prevalent among males than females at all frequency levels.

Those classified in a high-frequency firesetting group (those who has set 12+ fires in the last 12 months) were more likely than nonfiresetters to report low parental monitoring, high sensation seeking, delinquent acts, binge drinking, frequent cannabis use, use of illicit drugs, elevated psychological distress, and suicidal intent. Furthermore, the number of risk factors increased incrementally according to firesetting severity. That is, across the firesetting groups, the percentage of youth who set a fire during the past year was 12.6% for those with no risk factors; 26% for those with one risk factor; 34.9% among those with two risk factors; and 53.2% for youth with three or more risk factors.

Parental correlates of firesetting have included heightened personal distress, marital discord, and exposure to stressful life events, and less child acceptance, monitoring, discipline, and involvement in activities promoting the child's personal and family development (Kolko & Kazdin, 1986, 1991a). Firesetters have characterized their parent's child rearing practices as involving greater lax discipline, nonenforcement, and anxiety induction, with scores for matchplayers generally falling between firesetters and nonfiresetters.

Subgroups or Types of Firesetters

Efforts to understand the interrelationships among various fire-specific and general clinical characteristics have led to the postulation of subtypes of firesetters over several decades. Early subtypes identified several subgroups (e.g., curious, cry-for-help, pathologic, delinquent). Some children have engaged in serious and intentional fires, as well as concealed and destructive behaviors (Bumpass, Fagelman, & Brix, 1983; Jacobson, 1985; Kuhnley, Hendren, & Quinlan, 1982; Lewis, 1951; Stewart & Culver, 1982); others have set a single fire at home, which appeared accidental and due to curiosity or experimentation (Lewis, 1951). Unfortunately, there is no clear scientific support for any of these subgroups, especially since few validation studies have been reported (Kolko & Foster, in press). Still, it is important to consider how various aspects of a child's repertoire might be related and in what ways intervention may be facilitated by knowing these patterns.

Assessment and Evaluation Tools

Screening and assessment methods for use with children who set fires and their caregivers have been reported earlier and are more fully described in several sources (see Kolko, 2002d; Wilcox & Kolko, 2002). Most involve the use of rating scales, interviews, and either observational measures or incident reports based on the fire incident. Assessment should be tailored to the population and the context in which the problem is documented and managed. Of course, the population of children who set fires can be quite heterogeneous (National Association of State Fire Marshals, 1999). For example, there is much diversity in the types and severity of mental health problems reported among these children (e.g., aggression, impulsivity), caregivers (e.g., psychiatric distress, poor monitoring), and families (e.g. conflict, few rules; see Faranda, Katskikas, Lim, & Fegley, 2007; Kolko, 2002c; Massachusetts Coalition for Juvenile Firesetter Intervention Programs, 2002). Some young firesetters present to community agencies with complex psychosocial difficulties and involvement with several agencies or systems (Massachusetts Coalition for Juvenile Firesetter Intervention Programs, 2004). Consequently, a combination of interventions may be necessary to adequately address their needs (Massachusetts Coalition for Juvenile Firesetter Intervention Programs, 2004), as discussed further in a later section.

In general, most of the intervention programs in the United States (Kolko et al., 2008) and Canada (Mackay, Ruttle, & Ward, 2012) are affiliated with the fire service and provide some form of fire safety education. Some of them offer formal clinical services through the program or coordinated referral to a local provider. A comprehensive program includes resources to assess child and family variables associated with firesetting, teach fire safety skills, and offer psychosocial interventions to both children and their caregivers.

To address a child's exposure to and interest in fire, fire safety education (FSE) often provides training in safety skills/practices (Federal Emergency Management Agency, 1983a, b; see Pinsonneault, 2002). Among these educational materials, there is The Learn Not To Burn program, (National Fire Protection Association, 1979), which teaches protection (e.g., fire drills), prevention (e.g., using matches safely), and persuasion (e.g., practicing safe smoking). Even after two decades, this program is one of the few programs that has demonstrated its impact on fire safety knowledge, relative to controls (National Fire Protection Association, 1978). There are also benefits associated with other skills training procedures (Jones, Kazdin, & Haney, 1981; Jones, Ollendick, & Shinske, 1989) and more comprehensive fire safety curricula (Pinsonneault, 2002).

A comparison study with young hospitalized firesetters found that group fire safety/prevention skills training (FSST) was more effective than individual fire awareness/discussion (FAD) in reducing contact with fire-related materials, increasing fire safety knowledge, and reducing any follow-up fireplay (Kolko, Watson, & Faust, 1991). FSST (vs. FAD) children engaged less often in all four forms of involvement with fire (16.7% vs. 58.3%). In a residential treatment setting, one of 35 children exposed to a fire safety education (FSE) program had set another fire by 1-year follow-up (DeSalvatore & Hornstein, 1991).

Treatment applications designed to alter behavioral dysfunction and environmental conditions have incorporated several cognitive-behavioral therapy (CBT) procedures, at times, in conjunction with fire safety training (Kolko, 2002b). These methods include graphs that depict the antecedents of and consequences following a fire (Bumpass, Brix, & Preston, 1985), social/assertion skills training (DeSalvatore & Hornstein, 1991; Kolko & Ammerman, 1988; McGrath, Marshall, & Prior, 1979), contingency management (Adler, Nunn, Northam, Lebnan, & Ross, 1994; Kolko, 1983), and both parental and medication treatment (Cox-Jones, Lubetsky, Fultz, & Kolko, 1990). In general, treatment has emphasized child self-control, parent management skills, and/or positive family interactions, with some interventions incorporating several procedures (e.g., contingencies and/or behavioral training skills). These case study reports note reduced firesetting during follow-up and suggest the need to consider the behavioral and functional context of a fire, the child's interpersonal repertoire, and the use of effective parenting practices/consequences.

An Australian study evaluated similar approaches to working with firesetters (5–16 yrs.) classified as either curiosity or "pathological" (Adler et al., 1994). Although there was a reduction in the number of fires reported after intervention, there were no significant effects for the combined experimental interventions that were used with either firesetter group. Curious (vs. pathological) cases tended to show greater improvement (73% vs. 52%) and a lower percentage of dropout (20% vs. 35%).

In a prospective clinical trial (Kolko, 2001), firesetting boys and their parents were randomly assigned to one of two conditions that reflect contemporary methods used by community-based FSE programs and mental health clinics (CBT). Each intervention was compared with a brief condition that was already being used by the fire service (home visit from a firefighter or HVF). Although all three conditions showed improvements at posttreatment on measures of fire involvement, interest, and risk, CBT and FSE were more efficacious (than HVF) in reducing the frequency of firesetting and proportion of children who later played with matches, the severity of individualized child problems with fire, and the child's involvement in fire-related acts and other deviant fire activities. These and other group differences, along with certain time effects, remained significant at 1-year follow-up.

A follow-up report from that study examined the specificity, potential moderators, and predictors of recidivism after exposure to these three intervention conditions (Kolko, Herschell, & Scharf, 2006). In terms of specificity, FSE exerted specific effects on some fire knowledge and safety measures, as expected, and CBT tended to show specific effects on success in generating positive problem solutions. Potential moderators of outcome were suggested in exploratory analyses for FSE (i.e., exposure to fire models/materials, child's general fire knowledge) and CBT (i.e., family dysfunction). Fire history, fire attraction, and externalizing behaviors were among the predictors of firesetting recidivism. Thus, this information suggests that the two primary interventions have some specific effects and that certain background characteristics may enhance intervention response. These overall observations suggest the role of FSE (to target experience with, exposure to, and interest in fire) and psychosocial intervention to promote behavioral control to target individual/family conditions that influence child behavior (Kolko, 2002b).

Finally, an intervention program in New Zealand (Fire Awareness and Intervention Program—FAIP) bears mention (Lambie, Ioane, Randell, & Seymour, 2013). The FAIP is a nationwide educational program established in 1992 and available to children and adolescents up to the age of 18 years who have engaged in fire-related behaviors. The FAIP is delivered in a total of five regions across New Zealand and provides intervention for children from the ages of 3 to 17 years, and their parents. Delivered by fire service personnel in the child's home setting, the FAIP program has an advisory psychologist who acts in a supporting role for practitioners and provides training, advice on referrals to other agencies, and regular supervision to practitioners. The consultant psychologist also undertakes assessments where a practitioner identifies a child as being at high risk of further firesetting.

Surveys of Community-Based Intervention Programs

Descriptions of the characteristics, functions, and service delivery issues associated with community treatment have been reported for several decades (Kolko, 1988, 2002d). More recently, there has been significant growth in the availability and scope of community-based intervention programs and materials to educate and treat juvenile firesetters (see Faranda, Katskikas, & Lim, 2001; Henderson, MacKay, & Peterson-Badali, 2006; Okulitch & Pinsonneault, 2002). Some programs have organized into statewide coalitions, consisting of several affiliated program sites. Thus far, large-scale coalitions were found only in Massachusetts (Massachusetts Coalition for Juvenile Firesetter Intervention Programs, 2001) and Oregon (Oregon Office of State Fire Marshal, 2002a). Both coalitions are based on a set of common operating principles, procedures, staff, and resources. The model in Massachusetts embraces eight principles, such as the use of regional multidisciplinary task forces, formalized interagency agreements, early identification and referral pathways, and clinical assessment for high risk cases. The model in Oregon emphasizes the importance of community fire safety, program accountability, a continuum of assessment and intervention services, competency-based skills training, and local partnerships.

A survey of juvenile firesetter intervention programs was conducted to empirically depict the "state of the art" in serving this population in North America (Kolko et al., 2008). A total of 150 programs completed surveys, of which 31 were affiliated with one of these two state-wide coalitions. Survey questions examined staff and program composition, child and family characteristics, program operations, and needed developments. A total of 8,501 children and youth were served by these programs. Results indicated that most were males between the ages of 7 and 12 years. Overt behavioral difficulties, hyperactivity, peer problems, and rates of physical and sexual abuse that paralleled national rates were reported.

Paralleling the diverse clinical problems reported among child firesetters, nearly two thirds of parents were perceived as having multiple stressors or problems, including marital/partner problems, poor understanding of fire safety/prevention, substance use problems, use of harsh discipline and poor parenting practices, and a lack of involvement with the child, similar to reports from other programs (Massachusetts Coalition for Juvenile Firesetter Intervention Programs, 2002, 2004). These characteristics are similar to those described in other reports of juvenile firesetters (Faranda et al., 2007; Kolko, 2002c; Kolko & Kazdin, 1986).

Most of the programs (about 75%) were housed in the fire service and about 59% of all staff were fire service professionals, whereas the remaining 25% of the programs were housed in other service systems and included staff representing several disciplines (e.g., law enforcement, mental health, social services). The most common service involved child or parent interview (80% and 90%, respectively). Among intervention components, FSE was the most common. About one half of the programs offered brief counseling for both children and parents, and had follow-ups with a professional in the program, with less than one third offering more extended services. This modest rate of intervention services not only is in accord with the availability of mental health resources, but also highlights recognition of the need to address the child's and parent's mental health problems (Kolko, 2002a). Thus, for certain cases, services that extend beyond assessment and single-session fire-safety education will be necessary.

The average recidivism rate for all participants (both those who were referred back to the program for services and those who were not) was about 3%, although the range across programs was substantial (see Faranda et al., 2001). This rate is consistent with prior surveys (Kolko & Kazdin, 1986), but is far lower than the rates reported in empirical studies in which both children and their parents have been prospectively surveyed.

Finally, comparisons between programs that were or were not affiliated with a state-wide coalition revealed many similarities between programs (e.g., program size, number of referrals and ages, training, program type), but also some differences, with the coalition-affiliated programs reporting greater access to multidisciplinary resources, use of teaching and evaluation of fire science classes, and use of structured protocols for service provision. Coalition-affiliated programs also reported treating a population with higher rates of intentional firesetting, and individual, parent, and family mental health problems. Overall, these results highlight considerable diversity among the programs' multidisciplinary composition, assessment strategies, education and treatment services, and child and family populations. Comparisons between state coalition and independent programs revealed many similarities (e.g., program size, number of referrals and ages, training, program type), and a few key differences in methods, infrastructure, and philosophy.

Few surveys of mental health professionals have been reported. One report of psychotherapists was directed toward understanding some of the training and practice needs of mental health practitioners. The survey confirmed that practitioners as a whole have: (1) a poor awareness of juvenile firesetting behavior, (2) limited interest in the issue, and (3) a lack of knowledge of the current literature and available resources (National Association of State Fire Marshals, 2001). This book is, in part, designed to address this limitation.

Recidivism

Recidivism is somewhat common among children who have set fires, even after intervention (see Kolko, 2002b; Oregon Office of State Fire Marshal, 2002b). A prospective, non-intervention study that followed a sample of 138 children for one year, showed that 14 of 78 (18%) nonfiresetters—defined as children who had no history of either firesetting or matchplay—later did set a fire, and that 21 of 60 firesetters (35%) set an additional fire by follow-up (Kolko & Kazdin, 1992). Late starting was associated only with limited family sociability, whereas recidivism was associated with child knowledge about combustibles and involvement in fire-related activities, parent reports of receiving complaints from others in the community about the child's fire-related behavior, child hostility, lax discipline, family conflict, and limited parental acceptance, family affiliation, and organization. Some of these variables parallel those that have been associated with adult arson (Rice & Harris, 1991).

A 2-year follow-up study of 268 patient and nonpatient children (ages 6–13 yrs) that included some of the sample reported previously used fire history reports to classify cases into one of three mutually exclusive categories in order to determine how many children engaged in firesetting or matchplay only (Kolko, Day, Bridge, & Kazdin, 2001). Based on the aggregated reports of children and their parents, both patients and nonpatients reported high levels of follow-up firesetting (49%, 64%) and matchplay (57%, 76%), though the frequency of each behavior was generally higher for patients than nonpatients for both firesetting (*M*'s = 4.2 vs. 1.0) and matchplay (*M*'s = 3.1 vs. 0.9), respectively. In these samples, 25 of 50 nonpatients (50%) and 26 of 44 patients (59%) were recidivists, whereas 14 of 110 nonpatients (13%) and 11 of 42 patients (26%) became late starters. Other reports from inpatient or forensic settings noted 1-year follow-up recidivism rates of 44% (Stewart & Culver, 1982), 9% (Strachan, 1981), and 15% (Repo, Virkkuen, Rawlings, & Linnoila, 1997). Such findings highlight the prevalence of firesetting in clinic and nonclinic samples, and the continuity of firesetting over time.

The Rochester survey noted earlier included a follow-up survey of students who were involved in local fire department services, which reported a low rate of 3% recidivism. Further, another intervention study (Kolko, 2001) reported 1-year recidivism rates for three interventions, namely, FSE (24%), CBT (22%), and an HVF (44%). These results highlight the discrepancies that can occur between retrospective and prospective data, but the extremely low average recidivism rate reported in the Rochester survey provides an indication of the potential benefits of these existing community services.

Few studies have evaluated predictors of recidivism among juvenile firesetters (Kolko et al., 2001, 2006; MacKay et al., 2006). In one prospective study with a 2-year follow-up period (Kolko et al., 2001), there were two predictors of recidivism among nonpatients (firesetting, covert antisocial behavior) and two for patients (matchplay, covert antisocial behavior). In an intervention trial, several variables were significantly related to recidivism documented 1 year after intervention (i.e., initial number of matchplay incidents, initial number of firesetting incidents, curiosity about fire, fire attraction, involvement in fire-related acts, severity of child behavior problems; Kolko et al., 2006). A longitudinal study of a large cohort of children referred for firesetting behavior found that fire interest and antisocial behavior were related, and that fire interest added to the prediction of recidivism beyond covert antisocial behavior alone (MacKay et al., 2006). A review of eight studies (Kennedy, Vale, Khan, & McAnaney, 2006) also identified several related variables that were associated with recidivism (prior involvement in firesetting, interest in fire, covert antisocial behavior, male, older age, poorer social skills, family dysfunction).

A recent study by Lambie, Ioane, Randell & Seymour (2013) assessed arson recidivism and other offending rates for a group of 182 firesetting children and adolescents referred to the New Zealand Fire Awareness and Intervention Program (FAIP) over a follow-up period of 10 years, concluded many firesetters are at risk of further delinquent activities, including severe offences. The study, mostly composed of male and intentional firesetters, aimed to investigate predictors of offending behaviors as well as variables associated with previous involvement in firesetting behavior and offending severity. Although the arson recidivism rate was low (2%), the overall rate of general offending was high (59%) during the follow-up period. Fifteen percent of the sample was classified as severe offenders (e.g., robbery, sexual assault, serious assault), 40% as moderate (e.g., arson, burglary or speeding offences), and 4% as minor (e.g., theft). Of the offenders, 12.6% had been imprisoned during the follow-up period. Offending was predicted by an experience of abuse and a previous firesetting behavior at the time of the FAIP intervention. The presence of family stress and a diagnosis of Attention Deficit Disorder or Attention Deficit Hyperactivity Disorder were associated with previous firesetting behavior. In addition, involvement with family violence (as a perpetrator, complainant, or victim) was associated with more severe offending behavior.

The emergence of programs that offer an array of services is an important development. This book draws upon experiences conducting treatment and intervention research, and collaboration with community programs. Clearly, having access to more information (National Association of State Fire Marshals, 2001; National Fire Protection Association, 1995) about the characteristics of firesetters: the nature, severity, and history of their firesetting experiences; existing program services and operations; program outcomes; and recommendations will enhanced service delivery.

It is also important to consider whether services for firesetters are more efficient or effective when implemented by an independent program or using a multi- or interdisciplinary approach (Okulitch & Pinsonneault, 2002). The development of multidisciplinary teams that work with juvenile firesetters has been advocated extensively (see Okulitch & Pinsonneault, 2002), especially given the often diverse and complicated needs of these children and their families. This includes the various clinical problems reported for children, parents, and families, which supports targeting parental adjustment problems and parenting practices. Although the parent/family context in which firesetting emerges is addressed by some programs, this context may warrant even greater attention in an effort to assist the whole family. Some of the components worth integrating in a broad program are reviewed in this book.

We also include some attention to program monitoring and evaluation, given that only one half of the programs surveyed earlier conducted some type of "pre-post" or formal evaluation of their effects. The inclusion of outcome measures is essential for accurate program evaluation. Certainly further empirical data on program involvement and recidivism from local programs are needed (see Pinsonneault, 2002).

Because of the prevalence of fire safety education, we include some materials based on current educational curricula designed to enhance a child's understanding of the dangers of fire and ways to control or prevent fire. There are now many resources varying in content, developmental level, and focus (see Pinsonneault, 2002), though few have been formally evaluated.

In summary, there is a gradually accumulating literature that describes children who set fires, the fires they set, and the programs that serve them. We have outlined key clinical characteristics, services or operations, and likely outcomes. Key components include fire safety education and clinical intervention, which are extensively reviewed in the following chapters. We have also highlighted the need to conduct more formal follow-ups and work collaboratively. Such developments in program monitoring and coordination are necessary to maximize the efficiency, impact, and accountability of these services, thereby helping us to better identify and address the contributors to firesetting recidivism (see Kolko et al., 2006). The chapters that follow are designed to help the practitioner successfully integrate and then implement several assessment and interventions methods with children and youth who present with this troublesome and potentially dangerous behavior.

Overview of This Clinical Guide

This Session Guide describes a modular behavior intervention approach to working with children or adolescents with concerns about their interest in, exposure to, and/or use of fire. The approach incorporates content from prior intervention descriptions or outcome studies (Kolko, 2001; Kolko & Swenson, 2002; Kolko et al., 1991, 2009), as well as some materials or concepts adapted from other intervention programs for children with behavior problems (Kolko et al., 2009, 2012). Several of these methods were recently incorporated in an intervention model designed for families involved in conflict, anger, aggression, and physical force/abuse, called Alternatives for Families: A Cognitive Behavioral Therapy (AF-CBT; Kolko, Simonich, & Loiterstein, 2014).

Broadly conceived, the intervention material in this book is primarily directed toward reducing any inappropriate involvement in fire or related activities in order to lower the risk for causing any property damage, personal injury, or other adverse consequences associated with these behaviors. The material is most suitable for younger school-aged children with a recent history of fire involvement or use, but has also been adapted for use with older adolescents. Children or adolescents and their caregivers are the primary participants in treatment.

The intent of the Guide is to present a model that practitioners can use to provide comprehensive, evidence-based practice. The treatment proposed in this model generally targets various factors or domains that may encourage more appropriate and safe behaviors in both children/adolescents and their caregivers, and that provide skills likely to discourage fire involvement. In addition, the materials promote controlling exposure to incendiary materials and opportunities to use fire. We have tried to incorporate clinical techniques having at least some empirical support in the research arena and utility in the practice world.

What Is Included in This Guide?

This Guide and its incorporation of AF-CBT methods incorporates principles and techniques from several perspectives, including behavioral theory, family-systems models, and cognitive behavioral therapy, and lessons learned from the application of interventions in collaboration with fire service experts. The skills promote alternative interests, self-control and assertion, appropriate problem-solving, and communication, among other targets. The intervention methods are applied with an individual child/adolescent or caregiver, or jointly, to address both the firesetting behavior per se and the larger family context in which this activity occurs.

This approach seeks to promote the expression of safe and prosocial behavior, while discouraging any unsanctioned involvement with fire, by focusing upon instruction in intrapersonal and interpersonal skills. To do so, the intervention targets potential contributors to fire use, as well as some common characteristics of, or consequences exhibited by, children/adolescents and families after a fire. Potential contributors include strong interest in or attraction to fire, exposure to inappropriate materials, poor supervision or monitoring of children/adolescents, and harsh parenting practices, as well as coercive family interactions and heightened stressful life events. Potential consequences of fire involvement include worry and uncertainly about potentially imminent dangerous behavior, extreme changes in parenting and family activities, and family conflict.

Implementation: Feedback and Guidelines

OVERVIEW

This Guide presents strategies and resources for offering evidence-informed assessment and treatment for children and adolescents who have been involved in one or more fire-related incidents. This approach incorporates some adapted content from an intervention model called Alternatives for Families: A Cognitive Behavioral Therapy, (Kolko & Swenson, 2002; Kolko et al., 2014), as well as some unique materials or concepts developed for this population (Kolko & Foster, in press) or other populations referred for challenging behaviors (Fleischman et al., 1983; Robin & Foster, 1989). Broadly conceived, this assessment and treatment approach is primarily directed toward reducing all inappropriate involvement in fire or related activities in order to lower the risk for initial or continued property damage, personal injury, or other adverse consequences associated with these behaviors. This material is suitable for school-aged children and adolescents (generally 6 to 15 years of age) with a recent history of fire involvement or use. Children/adolescents and caregivers are both important participants in treatment.

The intent of the Guide is to present a model that practitioner can use with fire-involved children and adolescents as a component of comprehensive, evidence-based practice. The assessment and treatment strategies in this model generally target various factors or domains that may encourage more appropriate and safe behaviors in both children/adolescents and their caregivers, and that provide skills likely to discourage fire involvement. In addition, the materials promote controlling exposure to incendiary materials and opportunities to use fire. We have incorporated clinical principles and techniques from several perspectives, including behavioral theory, family-systems models, and cognitive behavioral therapy (CBT). Most importantly, we have also incorporated "lessons learned" from the application of these interventions in collaboration with fire service experts and mental health practitioners (Box 2.1).

Many skills taught by practitioners to fire-involved children promote alternative/positive interests/activities, self-control and assertion, appropriate problem solving, and communication. The intervention includes methods applied on an individual child/adolescent, individual caregiver, and a joint basis to address both the firesetting behavior and the larger familial context in which this activity often occurs. This approach seeks to promote the expression of safe and prosocial behavior, while discouraging any unsanctioned involvement with fire by focusing upon instruction on intrapersonal and interpersonal skills. To do so, the intervention targets

Box 2.1 Summary of Recommendations for Implementation Based on Efficacy and Field Trials with Juvenile Firesetters

- Collaborate with community partners (fire departments, law enforcement, court services, schools, social services) to include a mental health specialist in plans to assess whether family or behavioral issues contribute to fire incidents involving children and adolescents.
- Include authorities in referral and service plans to promote family participation in assessment and treatment of fire-involved youth.
- Use evidence-based strategies to understand and address fire involvement.
- Include both the fire-involved child/adolescent and a caregiver in the assessment and treatment process.
- Seek training and case consultation from mental health professionals who have experience and expertise in working with fire-involved youth.
- Use strategies and resources in this Guide flexibly based on the needs of individual children and families.
- Recognize the need for more intensive services (e.g., medication management, case management, support in multiple contexts) for some fire-involved youth and facilitate access to these when indicated.

potential contributors to fire use as well as common characteristics of or consequences experienced by children/adolescents and their families after a fire. Potential contributors include strong interest in or attraction to fire, exposure to individuals who use fire or to firesetting materials, poor supervision or monitoring of children/adolescents, and harsh parenting practices, as well as coercive family interactions and heightened stressful life events. Potential consequences of fire involvement include worry and uncertainty about potentially imminent dangerous behavior, extreme changes in parenting and family activities, and family conflict.

Who Should Use this Guide?

This Guide is intended to be implemented primarily by mental health professionals who have a masters-degree level (or higher) degree education, training, and licensure (or certification) and expertise in providing mental or behavioral health services for children, adolescents, and their families. But, it is also designed to support other practitioners involved with this population working in such systems as the fire service, social services, juvenile justice, and education. Within our experience in developing, implementing, and refining this Guide, we have worked collaboratively with mental health service providers from diverse training backgrounds, including clinical social work, clinical psychology, mental health counseling, counseling psychology, and psychiatry, as well as the aforementioned other practitioners. Because the professional terminology and core concepts of these training perspectives vary to some extent, we attempt to be very descriptive (and jargon-free) in outlining the objectives and activities for each component of the Guide. Nonetheless, our own professional training backgrounds are rooted in behavioral, cognitive-behavioral, and social-ecological perspectives on clinical child and adolescent psychology and the language we use in this Guide inevitably reflects this orientation.

Lessons Learned from Efficacy Research, Case Consultation, and Implementation Projects

The Guide integrates a broad range of research and practice experiences (e.g., clinical consultations, case studies, group comparison and clinical trials, field trials). The few early intervention studies conducted by the first author that were briefly described in the previous chapter reflected some general assumptions about the core elements that might best reduce the likelihood of a child's misuse of fire. This earlier work led to the development of a treatment manual that was used in a randomized controlled trial (Kolko & Kazdin, 1989).

Importantly, this treatment manual was refined and field tested in part through an implementation project carried out from 2009 through 2014 in a partnership between the Clinical Child Psychology Program at University of Kansas, the Family Conservancy (a nonprofit agency that provides mental health services in the Kansas City metropolitan area), and the Heart of America Fire Chiefs Association, which represents fire departments and community agencies that serve nearly two million residents in an eight county, bi-state metropolitan area. This partnership was supported in part by Fire Prevention and Safety grants from FEMA (EMW-2008-FP-02405, EMW-2011-FP-01327). This forum allowed for collaboration between the developer of the original intervention manual and assessment materials (DJK) and the director of this implementation project (EMV). Through this implementation project, we were able to work with community-based mental health practitioners as they used the assessment and intervention materials presented here with children and adolescents who were referred by fire departments, court services, schools, and other sources because of their involvement in one or more fire incidents.

The project allowed us to offer regular case consultation with Family Conservancy practitioners as they moved through the assessment and treatment process. We sought feedback on multiple occasions from the practitioners on the training and implementation process and used this feedback to improve the materials used in the Guide. We also gained valuable experience in working with fire departments and many other community agencies on the process of referring and connecting with fire-involved youth and their families for assessment and treatment.

The lessons we have learned from this broad range of experience are infused throughout this Guide. We summarize these "lessons learned" in Box 2.1 and provide examples in the following sections.

Lesson One

First, it is important to keep in mind that a child who misuses fire can be a danger to himself and to others, whether or not the child has any intent to cause a fire. One implication of this realization is that you need to recognize, understand, and at times communicate any of the child's high-risk behaviors and take the necessary precautions to lower these risks. This awareness should be informed by an understanding of any relevant ethical or legal issues or regulations invoked by the case (e.g., confidentiality, safety planning, duty to warn, danger to self or others). A fire can be a life and death situation, even though most of them do not result in any serious injuries or fatalities. But, some do.

Lesson Two

Similarly, we have included some fire-related assessment tools in this Guide. Most of them are used at intake to provide guidance on treatment selection and at discharge to provide some estimate of improvement. We encourage collection of this information at both times. But, it is also important to monitor the child's exposure to high-risk situations and involvement in high-risk behaviors on a continuous basis, in order to appreciate when additional supports or resources are needed to promote safety. So, for example, it is reasonable to consider asking the caregiver (or another adult with whom the child spends much time) to check the child's pockets, other possessions, or room for incendiaries or other forms of contraband, or to monitor his involvement with others who may have these in their possession. Certainly, explaining this procedure and then conducting this search in a sensitive and matter-of-fact way are important determinants of the utility and acceptability of these routines.

Lesson Three

Third, part of the success of our intervention trial was due to having developed a long-standing relationship with the local fire service. Even if you're not doing an intervention trial, it is important for you to collaborate with the fire service because these cases may require other services that help you focus on assessment and treatment, knowing that other systems are targeting key child or family problems, including social services, juvenile justice or law enforcement, education, or public health. In addition, the fire service is likely to be the primary referral source for cases with mental health problems that require specialty services.

Lesson Four

We have also learned that caregivers and their children often seem to need some of the same clinical content, including training in practical psychosocial self and other management skills. So, we sometimes teach children's self-regulation strategies to their caregivers to enhance parental competence and control. This is likely to be your judgment call, especially when the case has made limited progress. To prevent limited progress from happening, it is important for you to assess caregiver concern and motivation regarding the child's misuse of fire, insofar as some caregivers minimize its severity or may even find it amusing. The failure to appreciate the potential severity or high-risk nature of this behavior is poignantly depicted in a cartoon published in the New Yorker Magazine a few decades ago. In that cartoon, two parents, vaguely resembling members of the Adaam's family, are peeking with amusement into a somewhat dark and cavernous living room while their two young children are vigorously stoking a roaring fire. The caption, reflecting a thought bubble from one of the parents, says, "The little dears. They still believe in Santa Claus." Clearly, what the parents may have perceived about their children's behavior as cute or fun would, in reality, be far more appropriately viewed as dangerous and worrisome. Sadly, many parents actually do minimize the severity of their child's behavior in the context of fire and fail to take the necessary precautions to discourage the behavior at an early stage. The somewhat supportive parental reaction characterized in this early cartoon may be one of the most salient obstacles to reducing the problem of child firesetting.

Lesson Five

Finally, we found that caregivers were much more likely to attend sessions on fire safety education with their children, even without an invitation to join the child in the session, than when caregivers were scheduled for their own treatment sessions. They appeared to welcome this educational material and the time spent together in a family session with mostly stimulating and informative lesson content that included videos and handouts. In contrast, the rate of attendance for parent-only CBT sessions was lower, and it was clear that participation in those sessions was far more uncomfortable. So for this reason, we have found it useful to do fire safety early in treatment and, when possible, to include both parent and child together.

The Inclusion of a Mental Health Perspective in the Assessment and Treatment of Fire-Involved Youth

Given additional information and tools for assessment and treatment of firesetting, mental health practitioners are potentially well suited to assess and address situational and personal factors that contribute to a specific incident (or repeated incidents) of fire involvement. Although many children and adolescents report some degree of fire interest and involvement when asked, much of this goes undetected by their parents. Even when detected, firesetting is seldom given as a reason for referral of children and adolescents in most mental health settings, and most mental health providers have little experience or training in assessing or treating fire interest or involvement. The reasons why a child or adolescent actually sets something on fire can vary widely, and there clearly are instances when the misuse of fire by children and adolescents is part of a broader constellation of behavioral or psychological problems, family relationship problems, or involvement with deviant peers.

Fire involvement of children and adolescents is most often detected by community services including fire departments, law enforcement, schools, and child protection agencies. However, many communities lack clear or consistent plans regarding when or how to assess the extent to which family or behavioral issues contribute to identified firesetting behaviors. There appears to be great variability across jurisdictions in the level of collaboration among fire departments, law enforcement, juvenile justice, schools, and social services when a child or adolescent has been involved in a fire incident. Referral to a mental health practitioner for assessment of behavioral, psychological, and contextual issues regarding the misuse of fire appears to occur infrequently in many communities.

Even when a referral is made, the likelihood that a family will participate in assessment or treatment appears to be fairly low unless participation is required by court services or child protective services. For example, we conducted an implementation project that offered low-cost community-based mental health assessment and treatment for fire-involved youth in a major metropolitan area. Of the 91 fire-involved youth referred to this project over a 2-year period by fire departments, 45 did not attend any treatment or assessment sessions due to disinterest or lack of follow-up, including 21 children who were 10 years old or younger. Six of these 45 participated in other services, but 39 received no known follow-up. Of the six counties in this major metropolitan area, only one county routinely routed referrals for juvenile firesetting through court services and provided monitoring for compliance. When court services were involved, 88% of fire-involved youth attended assessment and treatment services. When court services were not involved (e.g., fire department referred families directly to the community-based mental health program), only 36%

of fire-involved children/adolescents participated in assessment or treatment. As a result, we recommend including court services or other community services that have the capacity to require families to participate in assessment (and treatment, when indicated) when developing plans to offer community-based mental health services for fire-involved youth.

The Importance of Collaboration with Key Community Resources

Due to the nature of firesetting, treatment with children and adolescents involved in firesetting often involves collaboration with other community agencies, including juvenile justice services, schools, child protective services, or fire departments. When collaborating with these agencies, it is important to clearly define the expected length of treatment and what is (and is not) involved in the intervention. Children and adolescents involved in fire incidents often meet diagnostic criteria for one or more mental disorder, and referral for medication management should be pursued when indicated. The Guide is primarily intended for use in outpatient or community settings, although the components are likely to be helpful for practitioners working with fire-involved youth in non-clinical (e.g., fire service) or more restrictive (e.g., detention, residential treatment) settings. Service coordination is an important component of intervention approaches for fire-involved children and adolescents who require support in multiple contexts, such as at school, in the home, and in community.

The skills addressed in this Guide may be important components of a larger service plan, but it is also important to recognize the need for more intensive services and facilitate access, when indicated. One example of a service in which community collaboration is valuable relates to the need for fire safety education. The Guide includes some material on this topic, but that chapter is not meant to be a substitute for a full course of fire safety education. Indeed, one of the two main intervention conditions tested in our prior clinical trials (Kolko, 2001) was an eight-session fire safety curriculum. Although much of that information is summarized herein, it is possible that some families may benefit from more extensive fire safety education beyond a mental health intervention. It is completely appropriate to suggest that the family may want to meet with a fire safety educator for this purpose. Besides meeting with a fire safety educator, the family may benefit from exploring some of the many educational videos, books, and online media about fire safety that have been designed for parents and children. These may be even more helpful when a family can watch and discuss them together. If a family is interested in these resources, it may be best to watch these videos after treatment has begun and you have established rapport between the child/adolescent and you. In our experience, sometimes these videos overemphasize that firesetting is bad and, consequently, the child/adolescent avoids discussing fire and firesetting incidents. Thus, it is important that these videos are accompanied by a discussion with parents and/or you to promote open communication.

Training for Practitioners

One issue that comes up often in disseminating clinical guidelines, such as this one, to real world settings, involves the concept of a "treatment manual." Practitioners sometimes believe that treatment manuals restrict their creativity or independence, or that they are being told what to say or do. Thus, it is important to note that strategies and resources in this Guide are intended to provide information on skills and concepts that are often relevant for fire-involved youth and their families, but it is up to you to determine which specific skills or concepts should be addressed with specific clients. When examples are given on how to phrase a particular skill or concept, it is not necessary (or desirable) to use the exact wording, unless this is part of a specific assessment item. Rather, you should use your own words to teach or convey concepts and skills.

Developing the capacity to offer this assessment and treatment requires an initial investment of time and effort for training with practitioners to work with fire-involved children and adolescents. Experienced practitioners—those who have strong backgrounds in behavioral, cognitive-behavioral, and family systems approaches—may be able to incorporate this Guide into their practice independently. More typically, we have found that it is helpful for practitioners to receive formal training from an experienced practitioner who can provide an initial overview of the materials and to participate in case consultation calls when the use of these materials with fire-involved youth and their families can be reviewed. Those practitioners who are less familiar with CBT and the use of treatment manuals may require more time to feel comfortable using this material. Formal consultation gives an opportunity to discuss case conceptualization, formulation of a treatment plan based on an initial assessment, and strategies for selecting and using specific modules from the Guide.

Tools for Tracking Implementation

We have developed measures for practitioners to record their delivery of specific components of the Guide on a session-by-session basis. We have additionally developed measures to assess practitioners' satisfaction with training procedures, their knowledge and confidence in using the screening and intervention measures, and the perceived utility of these procedures in meeting the treatment needs of fire-involved children and adolescents. We recommend gathering parent ratings of improvement in both firesetting indicators and related behavioral issues identified in the initial client assessment.

A Hierarchy of Intervention Components or Strategies

The cases referred for clinical services are likely to vary widely in the severity of a child's firesetting behavior and the level of dysfunction presented by the child or family. Given this variability, we think that it is helpful to consider the different options or components that make up a continuum of care for the child who misuses fire. Depicted in Figure 2.1 this continuum might serve as a guide for prioritizing the order and types of intervention strategies or methods that are selected for a given case. For example, it seems universally important for any case to restrict the child's access to incendiaries which, if properly done, could effectively eliminate the problem since the child would not be able to cause

Environmental risks/exposure to models → Detectors

Fire Safety & Skills → Films

Child Behavior (problems, prosocial) → Skills

Parent/Family Problems → Consequences

Court → Charges, Diversion, Restitution

Placement → Detention, Removal

Figure 2.1

A hierarchy of intervention components or strategies.

a fire. Therefore, this is a likely early step in an intervention program. Teaching fire safety skills to the child and parent might also be viewed as a fairly generalizable strategy that would be relevant to most cases, though the amount of attention paid to this topic could vary across cases. For those cases with significant child or family dysfunction, skills training in psychological management skills with children and parents is a logical addition to this treatment regimen. Of course, some cases either have prior juvenile or adult court involvement due to their fires, possibly with charges pending, which may require collaboration with the court and service coordination, or they will become involved with the court if they commit further infractions. In these cases, court involvement can be helpful in enhancing motivation to comply with treatment and in refraining from illegal or dangerous activity. Finally, there may be some cases that require an additional level of care, such as hospitalization or residential placement, to address significant psychiatric and/or family dysfunction. Appreciating the potential utility of all of these components may help you conceptualize the likely resources needed for a given case.

Service Dose, Order, and Intensity—Balancing Flexibility with Fidelity

Session outlines are provided to guide you through relevant topics. Sessions should be followed that meet the needs of the family. For some families, this will mean implementing the majority of sessions; others will only require a few sessions from each phase of treatment. Likewise, the length of time spent on each topic will vary according to the family's needs—at times, you may find it necessary to review a given topic for additional emphasis or, alternatively, to skip over a section of content that does not seem necessary to present. In general, the order in which sessions appear is recommended, but not required.

The Guide contains multiple therapeutic techniques, in-session exercises, and homework assignments. These have been developed to make the most efficient use of the brief treatment time available. These structured activities are meant to be helpful suggestions, not required elements. Certainly, the number of sessions, the content in the sessions, the order of activities within sessions, and the pacing of the intervention should be adapted to parents' and children's' developmental capacities, life circumstances, and treatment needs. Similarly, if information was reviewed once, but the family seems to have particular difficulty with that material, go back to a session or topic and review it. We understand that certain topics may require substantial review with some families, but may only require a quick summary or check-in with other families. Flexibility in the use of these materials is always a good option when justified.

Your primary responsibility in using this Guide is to help each family meet its treatment goals by integrating your clinical skills and knowledge with the provided session outlines. This may mean modifying or suspending use of the Session Guide to appropriately address more immediate concerns, crisis situations, or therapeutic issues that arise within families. We also appreciate that the information in this Guide may not be sufficient for a given client. We recommend that practitioners obtain appropriate training and ongoing supervision to help them effectively apply the materials in this Guide with their clients.

Assessment

We encourage practitioners who use this Guide to conduct an initial or intake assessment that yields clinically relevant information that can help guide treatment. What we mean by assessment in this instance includes arson investigator reports and other legal documents, formal standardized instruments, clinical interviews, and ongoing observations conducted during treatment related to the course and/or outcome of firesetting behavior. Some assessments may be directed toward understanding individual characteristics; other assessments may be focused on family-system characteristics. Likewise, some information may reflect fire-specific topics (e.g., curiosity with fire, exposure to fire materials, knowledge about fire), while other information may reflect more general topics (supervision, discipline, social skills). It is important to consider all of these potentially useful sources of information in both selecting treatment components and in adapting treatment to the needs of the family. Assessment is also beneficial at the conclusion of services to document improvements on key targets and to identify areas for further clinical work. Some suggested fire-related measures are described in chapter 3.

This Guide also draws upon the use of functional analysis procedures, which appear in the first session. However, for some children/adolescents it may be useful to repeat this activity throughout treatment with different events (not just fire related ones). With younger children, it may be necessary to regraph the fire each time to encourage a discussion, while with older children/adolescents a simple conversation about it may be sufficient. In particular, this may be helpful during the thought-feeling-behavior connection session or during a lesson about problem solving. The advantage of using functional analysis is that it can help to identify both proximal and distal contributors to a given incident as well as various types of consequences that were (or were not) administered after the incident. Understanding the ABC's of functional analysis (i.e., antecedents; behaviors that include thoughts, feelings, and actions; and consequences) is an important basic skill that practitioners often need to efficiently and effectively complete treatment.

Referral Based on Fire-Incident versus Other Reasons

This Guide was originally developed for use with children or adolescents who were referred following an allegation or documented incident of firesetting or fireplay. This therapy is also encouraged for cases where there is considerable risk for such activity (e.g., a strong interest in fires) though it may not yet have occurred. Clearly, most of the content in this Guide is directed toward understanding and addressing a specific fire-referral incident. However, in those cases where there is no such incident, it may be useful to understand the nature of the risk factors that are present and to target them, if possible. In both cases, adaptations of the material may be useful to more carefully target the specific circumstances involved in a given referral.

Developmental Sensitivity/Context

As with all intervention models, adaptations should be made to the session outlines and materials to meet the developmental needs of children, adolescents, and parents. The Guide includes separate examples and materials for use with school-age children versus adolescents. Children or parents with limited intellectual functioning, or very young children, may require more simplified services or translations of some of the more complicated treatment concepts. Similarly, adolescents typically require a greater focus on personal responsibility, autonomy, and parent-adolescent negotiation. In addition, we discourage the use of shock or scare tactics (e.g., provocative statements or video's) due to their potential to traumatize the child and their relative ineffectiveness in other areas.

Home Practice

We have incorporated some home practice assignments in each session. The goals of home practice are:

1. To remind clients to use newly learned skills
2. To help clients apply the lessons learned in sessions at home (generalization to different contexts)
3. To allow clients to report back to the practitioner so that he or she can help problem solve and refine the skills or their application so they work well for families
4. To prepare parents and children/adolescents for the next session so that session time can be conducted more efficiently

It is important to first determine whether the parent or child/adolescent can adequately understand and handle the assignment. This is the reason we sometimes offer a choice of different activities, which you should consider carefully before making an assignment for home practice. It is helpful to consider a few initial questions in selecting a home practice assignment:

1. Is each participant (child/adolescent and parent) literate and able to read English comfortably?
2. Can each complete a worksheet without help?
3. Does each participant have time to fit an assignment into his or her life at this time?
4. How much of the assignment will each participant be expected to complete?

How to Use This Guide

We have prepared this Guide in a format that we hope will lend itself to efficient use in any given session by a broad range of practitioners varying in orientation, experience, and comfort using CBT-based techniques. This content includes suggested wording, if needed, for the purpose of providing a verbatim dialogue when presenting and discussing session content. Also included are examples of the key topics to be addressed, activities within each topic, relevant exercises or role-plays, caveats and general recommendations, and home practice assignments for the next session. We hope that this organization provides an efficient method for delivering this material.

Each chapter has a brief session outline that refers to the main topics presented in that section. Most of the chapters include materials for use with parents and/or children/adolescents during the session or for completion as a follow-up activity. The purpose of these materials is to provide families with a tangible and organized set of reference materials to review in between sessions and after the completion of the program. They are also designed to help make the material relevant, engaging, and fun.

Forms specifically for the practitioner appear in appendix A at the end of this Guide. Forms for children can be found in appendix B, and forms for parents are located in appendix C.

Screening and Assessment

OVERVIEW

This chapter provides a brief summary of tools that we have developed to assess child and adolescent involvement in fire-related activities. These include a brief fire history screening questionnaire, two versions of a fire-incident analysis interview (one to be completed by the child and the other by a caregiver), and two versions of a structured interview (one to be completed by the child and the other by a caregiver) to evaluate fire-specific and general risk factors that may contribute to the misuse of fire. All of these tools should be completed separately with the child/adolescent and a caregiver. Some parents may be able to complete their tools on their own; others may require some support or may benefit from a direct interview.

Fire-specific screening tools can be used to understand the severity of the child's fire history or clinical problems, which may help to determine whether the child should receive fire safety education or mental health intervention. Assessment tools can be used to identify intervention targets or outcomes worthy of documentation over time. We include here some examples that have been studied in research and clinical settings. These tools are intended to supplement more general clinical assessment procedures that are typically used for diagnostic and treatment planning.

The overall purpose of these screening and assessment tools is to allow practitioners to gain a more comprehensive picture of the Intervention needs of fire-involved youth. This helps the practitioner and the family to determine the range of services that may be needed and to set specific treatment objectives. In some instances, a child's interest and involvement in the misuse of fire may be a relatively circumscribed problem that occurs in the context of generally good psychological and behavioral adjustment. More often, the misuse of fire appears to occur in the context of other difficulties, such as behavior or emotional problems, family system stress, or adjustment difficulties at school or in the community. Many times, these difficulties can be addressed with outpatient services. A subset of fire-involved children and adolescents may also have serious emotional disturbances that require placement in more restrictive settings or involvement in more intensive service plans that involve multiple providers, such as medication management, family support, school-based interventions, and court services.

From our perspective, incorporating the fire-specific tools presented here with broader clinical assessment procedures enhances treatment planning by giving a realistic idea of the range of services that may be needed to make meaningful changes in the adjustment of fire-involved youth. It is beyond the purview of this book to cover clinical assessment more broadly. We refer readers to evidence-based guidelines for assessment and treatment of children and adolescents with specific presenting problems in addition to fire involvement (see http://effectivechildtherapy.fiu.edu/; http://www.samhsa.gov/nrepp).

It also seems prudent to explain to the child and parent any conditions affecting the degree to which their information will be kept confidential. In general, it may help to indicate that the information they discuss is generally confidential and is not shared with other participants or outside professionals. However, they should both know whether you have an obligation to report any and all details of the child's firesetting incidents or other consequences, who will get what information, and how that information will likely be used. This is especially important when the practitioner works in a larger system or with other service sectors. In addition, there may be other professional requirements governing a discussion of certain details and their potential release to address safety concerns to the child or other persons that the practitioner should explain.

Screening for Misuse of Fire among Referred Children and Adolescents

The original Firesetting History Screen (FHS) was a 14-question, structured form, which asks about incidents within two time frames—within the 12 months (current period) and more than 12 months ago (past period) (Kolko & Kazdin, 1988b). The original FHS can be used routinely in clinical practice as a brief screening tool for detecting fire interest and involvement among children and adolescents referred for clinical services (Morris, Wilcox, Rubens, & Vernberg, 2012). It provides a clear and succinct glimpse of a child's fire interest and behavior, and can be given as a questionnaire or structured interview. It offers a sense of how often fire-related behavior has occurred, what was burned, the severity, while also delineating a time frame during which these events transpired. The parent and child screens are pragmatic. They can be administered efficiently and provide a very basic useful measure at locations, such as a pediatric clinic, school, court clinic, or mental health center, depending on the needs of the case. For instance, employing the screen in a child and adolescent psychiatric unit has proved helpful in identifying children and adolescents for interventions who have been actively engaged in setting fires, but whose firesetting has not previously been identified (Wilcox, Murphy, & Goldstein, 2000).

An abbreviated version of the FHS includes both caregiver and child questions. The screen provides for "yes" and "no," or short-answer response choices for the following questions:

1. Does the child like fire?
2. Did the child play with matches?
3. Did the child burn something or set anything on fire?
4. When did this happen?
5. What was burned?
6. How many times did the child set a fire?
7. How serious were the damages?
8. Was the fire department called?

The screens do not rely solely on the report of the child or adolescent. Parents are also asked the same questions about the child's behavior in the parent version of the screen. A screen form with these questions is shown in Practitioner Form 3.1. (As noted in chapter 2, all Practitioner Forms appear in appendix A at the end of this Guide.)

Due to the direct nature of the questions posed in the screening, there is often good agreement between parents and children on the details of the history. Kolko and Kazdin (1988a) found relatively high agreement between parent reports and the reports of children when administering the screen in an inpatient setting (mean Kappa = .61, range = .43 to 1.00) and outpatient setting (mean Kappa = .71, range = .55 to .91). At the same time, in a clinical encounter, the agreement or lack of agreement between the answers provided by the parents and the youth may be important to examine. The FHS has been used to identify current and past firesetting in several other studies, and has been found to be related to different measures of firesetting involvement and interest (see Kolko, 1999).

Documenting Details of a Fire

Designed to parallel the FHS, Firesetting Incident Analysis (FIA) interviews can be used to provide a quantitative assessment of various parameters or details of a specific firesetting incident, especially the most serious firesetting incident (Kolko & Kazdin, 1991). Although it was initially designed to only be administered to parents, a shorter measure was developed for children and adolescents (Kolko & Kazdin, 1994).

The Firesetting Incident Analysis—Parent Version (FIA-P) (shown on Practitioner Form 3.2 in appendix A) consists of 50 questions administered to a caregiver. Four domains of questions are surveyed:

1. The first domain includes details or characteristics of a firesetting incident (e.g., how materials were obtained, site of fire, type of property damage).
2. The second domain includes ratings of the levels of behavioral and emotional correlates that were evident in the two weeks that preceded the fire (e.g., aggression/defiance, depression/withdrawal, rule violations).
3. The third domain reflects the child's motives for the fire based on questions that offer five response choices on a Likert scale (1 = not true at all; 5 = very much true). Factor analysis of 12 items yielded three general motive factors (Curiosity, Anger, Attention/Help-seeking) and

two other items (accident, peer pressure or destructiveness) designed to understand the presumed reason for the fire.

4. Finally, the fourth domain questions are directed toward examining any consequences following the fire, such as family/disciplinary (e.g., family discipline, child was talked to or counseled by someone outside of the family), financial (e.g., value of damages), medical (i.e., injury, death), legal (e.g., criminal record, removal from home), and social/peer (e.g., peer acceptance, peer rejection/avoidance).

Findings from a study evaluating this instrument showed that heightened (vs. low) curiosity was associated with greater fire involvement out of the house and less costly fire damages, a higher degree of overt and covert behavior, hostility, and earlier experiences with fire. Heightened (vs. low) anger was associated with greater aggression/defiance just prior to the fire, greater exposure to fire, more reports of community complaints about their fire-related behavior, more fire-related behavior in general, more deviant behavior just prior to the firesetting incident, milder punishments as a consequence for misbehavior, and more peer rejection following the fire.

The parallel, 21-item Firesetting Incident Analysis—Child Version (FIA-C) (shown on Practitioner Form 3.3 in appendix A) also solicits information about a firesetting incident (Kolko & Kazdin, 1994). The child is asked a number of questions about the characteristics and details of the fire, including such questions as where he got the materials to start the fire, was the fire planned, where the fire occurred, the extent of the damages, and whether the child was alone or not. The child is also asked about his motive in setting the fire, whether it was primarily out of curiosity and experimentation or anger and revenge. In addition, a variety of other questions ask the child about the consequences of the incident as well as his reaction to the fire (e.g., discipline, attention, projected impact of the incident on future firesetting). Thus, the FIAC provides a structured approach to understanding the child's reports to questions regarding the incident.

The FIA-C has been shown to highlight factors that can contribute to a child setting a fire again (Kolko & Kazdin, 1994). This measure has been able to identify that when parents respond with few or no consequences in response to a firesetting incident, there is a greater likelihood that the child will engage in future incidents of matchplay or firesetting. Similarly, if a child shows a positive reaction to the fire or even a neutral reaction with little or no remorse, there is a greater likelihood of future incidents of firesetting. Four of the fire characteristics predicted the overall severity of the child's involvement in fire at 2-year follow-up (i.e., fire out of home, acknowledgment of being likely to set another fire, a neutral/positive reaction to the fire, no parental response to the fire). Such instruments may facilitate a quantitative evaluation of the details of an individual incident.

Assessing Fire-Specific and General Risks for Treatment Planning

As noted earlier in this book, certain characteristics that may increase a child's risk for firesetting are fire specific in nature in that they relate directly to access to, interest in, and exposure to fire materials. Other more general factors potentially related to the emergence of firesetting relate to the child's or parent's functioning, or the family environment. Assessment of both fire-specific and general risks for fire involvement is very useful for treatment planning and overall case conceptualization. Separate standardized interview instruments have been developed for parents (the Firesetting Risk Interview or FRI; see Kolko & Kazdin, 1989a) and children (the Children's Firesetting Interview or CFI; see Kolko & Kazdin, 1989b). These can be used to evaluate several fire-specific characteristics that may bear implications for intervention.

The FRI and CFI are based on a risk-factor model of firesetting (Kolko & Kazdin, 1986, 1989a, 1989b). Items on the FRI and CFI were drawn from earlier FEMA interviews for firesetting (1979, 1983), and from early empirical literature on risk factors for fire involvement (Kolko, 1985, 1989). Based on major findings reported in the juvenile firesetting literature, the authors synthesized and integrated factors within three major domains that may increase a child's probable risk for firesetting behavior. The domains include learning experiences and cues, personal repertoire, and parent and family influences and stressors. Variables associated with both fire-specific and general factors have been evaluated in various instruments.

The FRI and CFI have been refined over time through clinical experience and research. The two instruments can be administered to address three important assessment goals:

1. The first involves identifying the child's individualized firesetting risks, along with some other clinical concerns included in these instruments.
2. The second is to determine targets for intervention.
3. The third is to provide the evaluator with some objective outcome measures that may be useful in depicting any changes in targeted outcomes.

The FRI and CFI are noteworthy for their use of structured, objective scoring procedures and for their being subjected to empirical evaluation. Still, it is important to point out that the FRI and CFI should be used in the context of a comprehensive fire assessment that allows the evaluator to identify and understand potential targets for intervention or treatment. The FRI and CFI are designed to contribute information related to some of these potential targets. These tools also highlight the need to consider both individual and environmental variables in the formulation, as child variables may be secondary relative to other parental or contextual variables. Of course, it is critical to create a supportive context for information gathering and service delivery, given that the utility of any of this information derives from having a cooperative and an accurate informant.

The FRI includes 86 items, which are divided into two primary domains (fire-specific, general). There are 15 individual structured scales. The FRI examines several factors specific to fire and more general factors. The factors and some information about them appear on Practitioner Form 3.4 in appendix A.

The FRI can be administered as a self-report or, as needed, an interview. The items are scored by simply summing items within a given domain to get a total score. Most of the items are multiple choice (1–5 pt. scales), with items on one subscale (parental preparedness) scored on a yes/no (1 vs. 0) basis.

Each of the subscales in the FRI was found to be internally consistent; that is, statistics examining each factor indicated that the items actually go together very well and appear to be reflecting a common dimension. Thus, the instrument has some degree of reliability of measurement. Next, scores on each dimension have been compared for firesetters and nonfiresetters. Compared with nonfiresetters, parents of firesetters acknowledged significantly higher scores on measures of firesetting contact (e.g., curiosity about fire, involvement in fire-related acts, exposure to peers/family fire models, complaints about fire behavior, early experiences with fire), general child/parent behavior (e.g., increased negative behavior), and family environment (e.g., greater use of harsh punishment, less effective mild punishment).

Only a few differences were found between patients and nonpatients on the FRI. (Note that *patients* were children with a recent history of outpatient or inpatient treatment. regardless of firesetteing status; *nonpatients* were those who did not have any treatment involvement in the past year.) Specifically, patients received scores indicating greater curiosity, early experiences, complaints, and less exposure to mild punishment (efficacy) than nonpatients. In terms of different motives, heightened (vs. low) curiosity was found to be associated with certain fire-specific factors (i.e., Curiousity, Exposure, Early Experiences). Other differences were associated with reports of heightened (vs. low) anger (i.e., Involvement, Complaints, Exposure, Knowledge, Parental Preparedness).

Scores on the FRI have been used to classify cases (Firesetting vs. no-Firesetting) into their proper groupings. The FRI correctly classified 81% of the cases, based primarily on four factors (Curiosity, Involvement, Complaints, Negative Behavior). A few of these subscales have actually been found in a follow-up study to predict recidivism in nonpatients (Exposure) and patients (Complaints, Efficacy of Harsh Punishment) 2 years later (Kolko & Kazdin, 1992).

In terms of the use of the FRI for individual assessment purposes, scores can be interpreted in two ways. First, one can compare the resulting scores to the normative data included in the original article to determine how a given child compares with other subgroups of children varying in firesetting status (firesetters, non-firesetters) and patient status (nonpatients, outpatients, inpatients). Tables in that article describe the means and standard deviations for each subscale (Kolko & Kazdin, 1989a).

The second approach is more idiographic and involves examining each subscale score to determine if it appears high or low relative to the range of possible scores. This "eyeball" approach is meant to identify any noteworthy scores that may reflect different patterns of fire-specific involvement or exposure on an individual or case-by-case basis. In general, some children receive very high scores on fire interest and exposure variables (e.g., curiosity, exposure to fire materials/models), whereas others receive very low scores on fire interest or exposure variables, but show high scores on the dimensions relating to the child's deviant behavior and exposure to harsh punishment.

The 46-item CFI includes a smaller set of six risk factors believed to be easiest for children to answer. This tool is shown on Practitioner Form 3.5 in appendix A. The subscales are: curiosity about fire; involvement in fire-related activities; knowledge about things that burn; fire competence; exposure to models/materials; and supervision/discipline. The items are assessed on similar structured scales using multiple choice (1–5) items and a few yes/no items. The CFI is administered in a similar way to the FRI, but it also includes the use of certain props (e.g., fake matchpack or phone) for use in a few role-plays. Scores are also calculated for each subscale, with correct answers assigned to the yes/no (score = 1 for each correct item).

There is very good internal consistency and test-rest reliability for the six factors. Relative to nonfiresetters, firesetters acknowledged more attraction to fire, past fireplay, family interest in fire, exposure to friends or family who smoke, and, somewhat surprisingly, knowledge of things that burn, but they tended to show less fire competence (skill) on role-plays than nonfiresetters. No differences were found on the supervision/discipline factor. In terms of patient differences, *patients* (vs. *non-patients)* had higher scores on the exposure and involvement in fire-related activities subscales. All six factors in the CFI correctly classified 71% of these cases. Two of the subscales (competence, exposure) were found to serve as predictors of recidivism 2 years later (Kolko & Kazdin, 1992).

The interpretation of scores on the CFI dimensions is similar to the methods used with the FRI. Accordingly, one can compare the scores to the normative data included in the original article for general comparison purposes (firesetters, non-firesetters). And, one can examine each subscale score to determine if it appears "noteworthy" relative to the possible range of scores for a given case. One of the benefits of having a few dimensions in common between the FRI and CFI is the ability to obtain two perspectives on the same dimension (e.g., curiosity about fire) and then examine any differences in parent and child perceptions as to the significance of subscale.

Brief Case Example

To provide an example of the use of these tools, we report some summary information for an 8-year-old boy referred for intervention. Table 3.1 shows the results of an intake and discharge assessment on the FHS, and Figures 3.1 shows the results based on the FRI. At intake, caregiver reports on the FHS confirmed the child's interest in fire, and indicated several fire (but no matchplay) incidents. The motives of all three fires were attributed to heightened anger. Caregiver ratings on the FRI indicated that the child had low levels of exposure to fire models/materials, low levels of involvement in fire-related activities, and low levels of negative behavior, was perceived as having a high level of fire knowledge, and was exposed to appropriate levels of punishment. However, he had high scores in fire curiosity and low scores in fire competency and parental preparedness to address fire at home.

This case might be typical of those children who could benefit from fire safety skills training.

Indeed, the child and parent did receive such training and the discharge assessment scores on the FHS and FRI are also indicated in Table 3.1 and Figure 3.1, respectively. The discharge scores suggest that there were some key improvements on targeted risk factors, notably a lower level of perceived fire curiosity or attraction and a higher level of perceived fire

Table 3.1 Firesetting History Screen Assessment Measures at Intake and Discharge for a Child Who Received Fire Safety Education

Firesetting History Screen Results

Items	Intake	Discharge
Interest	Yes	No
Fire incidents	3	0
Matchplay incidents	0	1
Seen with matches	1	1
Motive	Mad (3)	–

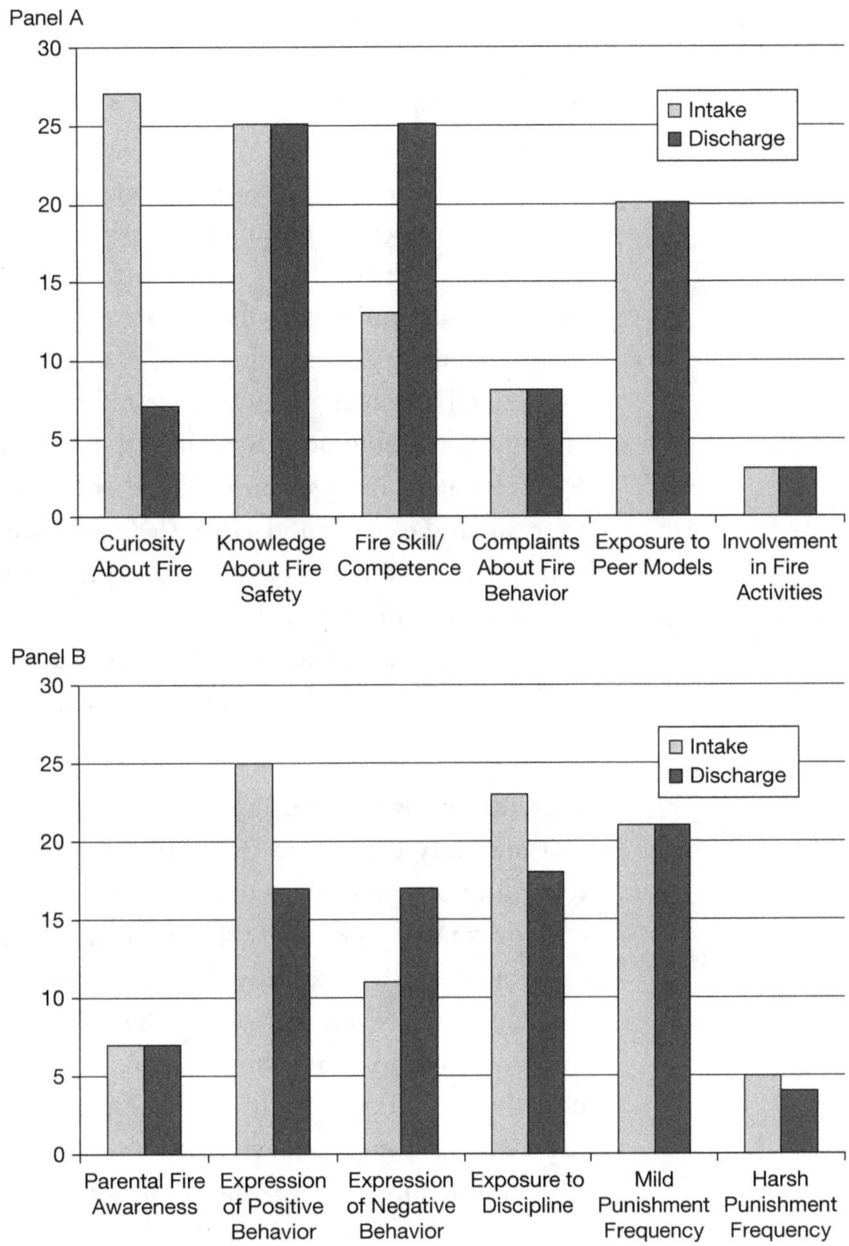

Figure 3.1

Firesetting risk inventory scores at intake and discharge for a child who received fire safety education.

competences/skill (Panel A). Interestingly, there was no change in several other risk factors and even a slight increase in the level of the child's negative behaviors (Panel B). Still, the practitioner and caregiver saw the child as being improved on key outcomes at the time they agreed to end services.

There is no simple way to document and understand all of the potential characteristics of a child's history of misusing fire or the most relevant characteristics of a child, his/her caregivers, the home or family environment, other social context information, and the family's intervention experiences. Clearly, this agenda could cover considerable ground, so it is helpful to use initial assessments to guide the selection of later ones. We suggest that you begin with a brief evaluation of history of firesetting to understand how many fires were set and to use this information to stimulate questions about other aspects of this history that might seem relevant (e.g., settings, motives, context). Other assessment tools can be used to learn about the child's and caregiver's repertoire in relation to fire, some of which may serve as risk factors for setting a fire. These factors include the child's interest in or attraction to fire, knowledge of things that burn, the level of supervision, the extent to which he/she has hidden or concealed his/her firesetting behavior, and his/her degree of exposure to fire models or material.

A child's motive for setting a fire can also be a challenge to identify and may vary one fire to the next. For some, the act of firesetting is a rewarding or enjoyable act; for others, it may be an attempt to influence or control a person or situation. Likewise, the possible lack of any consequences may also provide feedback to the child that may enhance the likelihood of another one. It helps to construct a history of these incidents to learn about the larger context that has enabled or even supported the child's use of fire and how it serves to convey the child's needs (e.g., curiosity, anger, relieve boredom, self-assertion with peers/parents). It is in the course of gathering this history of a child's firesetting behavior that we can begin to understand what has motivated and possibly reinforced or rewarded the child's continued interest in setting fires, while also gaining a better understanding of the role firesetting plays within the life of the child and family.

Understanding the Child's Firesetting

MATERIALS

- "Fire Graphing Sheet" (Practitioner Form 4.1)
- Markers or pens with three or four different colors
- Basic facts regarding the child's most recent firesetting incident and his most serious firesetting incident
- Rapport building games, such as cards, Jenga™, etc. (optional)
- Pencils, pens, and blank paper

OBJECTIVES

- Introduce the purpose of your work with the child (i.e., to reduce the risk of the child getting hurt by firesetting, or hurting others).
- Begin to develop rapport with the child.
- Obtain information regarding the child's ability to self-reflect and self-monitor during problem situations.
- Obtain information about behaviors and consequences regarding the fire, including emotional and cognitive causes.
- Obtain information about potential acceptance of responsibility for the fire.

Introduce the Purpose of the Intervention

In the first treatment session, it is important to introduce the purpose of your work to the child, which is to reduce the risk of the child getting hurt by firesetting, or hurting others. You may want to say something like the following:

We work with other children, like you, who have used fire or who just like fire for different reasons. One of the things that I would like to do now is to talk to you about your experiences with fire (i.e., what kinds of things you've done with fire, what you might like about fire). This is important so we can figure out ways to help you and your family stay safe. For those children who really like fire, we might help them find other ways to have fun. For those who use fire because it helps them get things they want, we try to help them get those things without having to do something that can be dangerous. We also try to help parents get along better with their children and to help families be more positive with and helpful to each other. Does that sound ok to you?

Practitioner Note

Try to deal with any noncompliance or apparent disinterest by asking the child to see what this will be like before she decides that she doesn't want to participate.

Build Rapport

Before discussing the child's problem behaviors, it is helpful to get to know the child by asking a few questions: about himself, what he does, about who's important in his life (friends, parents, siblings), and his past experiences with services (whether it was helpful, whether he liked it, etc.). Ask additional questions that would seem to relate to other answers given by the child or to other information you have about the child (e.g., hobbies, interests, skills) to establish some rapport and to elicit cooperation. If a child appears shy or not engaged, play games together to facilitate the development of a positive rapport.

Evaluate Child's General Capacity for Accurate Self-Reflection

Gauge the child's current skill in accurate self-reflection about problem situations in general. This helps you get an initial sense of the child's style of coping and responding to conflict/frustration. The discussion also may show how well the child might be able to understand the connection between specific feelings, self-statements, and behaviors, including those associated with firesetting.

Start your discussion by asking the following questions about problem situations:

- *You told me about things you like to do or are good at. Are there some things you don't like—what are they? Are there things you aren't good at—what are they?*
- *Do you ever find that things don't work out the way you want them to?*
- *Some of the children we see tell us that they sometimes don't get along with everyone—is that ever true for you—how?*
- *What kinds of things bother you or get you into trouble?*
- *How would you like things in your life to be different?*

Explain the Role of Feelings, Thoughts, and Behavior

When something happens to us we have three reactions.

- *One type of reaction is* feelings. *For example, if we get in a fight we might be feeling angry, which might make our heart beat more quickly, our fists clench, or our stomach might get upset.*
- *The second type of reaction is* thinking, *which is what we tell ourselves. Like if we were angry, we might think, "'I am going to explode" or "Why am I so stupid?"*
- *The third type of reaction is* behavior, *that is, what people do to try to feel better, like yelling, throwing a toy, or punching a wall.*

I'm going to ask you about each of these reactions as we talk about your fire (i.e., what you burned).

Solicit Key Contextual Details Using Basic Probe Questions

Try to use these probe questions to identify individual situations in which the child has experienced unpleasant or distressing emotions and any associated negative thoughts by asking questions like the following:

- *What happened (nature of the incident)?*
- *Who was involved?*
- *Why did it happen (initial reason for the incident)?*
- *What were you thinking about (or feeling) at the time (thoughts and feelings experienced during the incident)?*
- *What did you do during (and after) the incident?*
- *How did it end (outcome)?*

Graph a Fire Incident—Conducting a "Functional Analysis"

Sample Rationale for Making a Graph

Graphing consists of charting a particular event on a graph that represents time on one axis (horizontal) which can be populated with key events or behaviors and the intensity of an emotional state/reaction or specific thought on the other axis (vertical). Some background description of the use of this procedure and the importance of conducting a functional analysis and related methods to learn about the context of a child's firesetting incident can be found in our prior work (Kolko, 2002). Usually, the fire is placed somewhere near the middle of the horizontal axis. Once those events are included, changes in the child's feelings and thoughts are drawn where they are linked to the key activities or events listed on the axis, that is, those associated with the fire incident (or event) itself. Here, an attempt is made to review the available facts regarding the child's most recent incident with the goal of teaching the child to identify those incidents or situations that typically elicit the affective states and behaviors that may precede the use of fire. Those experiences that precede the fire reflect antecedents; those that follow the fire are called consequences.

Certainly the types of immediate antecedents to fires are quite variable, but common ones include events that trigger certain experiences. For example, some children who are left alone or who are simply understimulated may experience boredom or loneliness, which may lead them to use fire to have something to do for fun, especially when they happen to find a lighter in the house. Interpersonal conflicts (e.g., with peers, parents) are also common and may lead a child or adolescent to experience feelings of anger, frustration, or hostility that may serve as a motivation to "get even with" or control the other person. This motive is commonly resolved or mediated by setting the fire, which may reflect the desire to destroy something that belongs to the other person.

Common immediate consequences for setting a fire may include having to clean up the burned area, getting medical attention, being caught, or receiving a consequence from a caregiver following the fire. Of course, sometimes children experience more positive or benign consequences (e.g., peer attention, no response).

For illustration, the sample fire graph in Figure 4.1 below depicts the interplay among these variables that were related to a fire set by one adolescent boy with a history of multiple firesetting incidents. This fire occurred during a sleepover with some cousins. The boy said that they weren't having much fun and didn't have much to do. The fire was set after the cousins left the boy alone. During the session, the boy was asked several questions about what happened, how he felt and thought, and what else happened both before and after the fire. The child was asked to

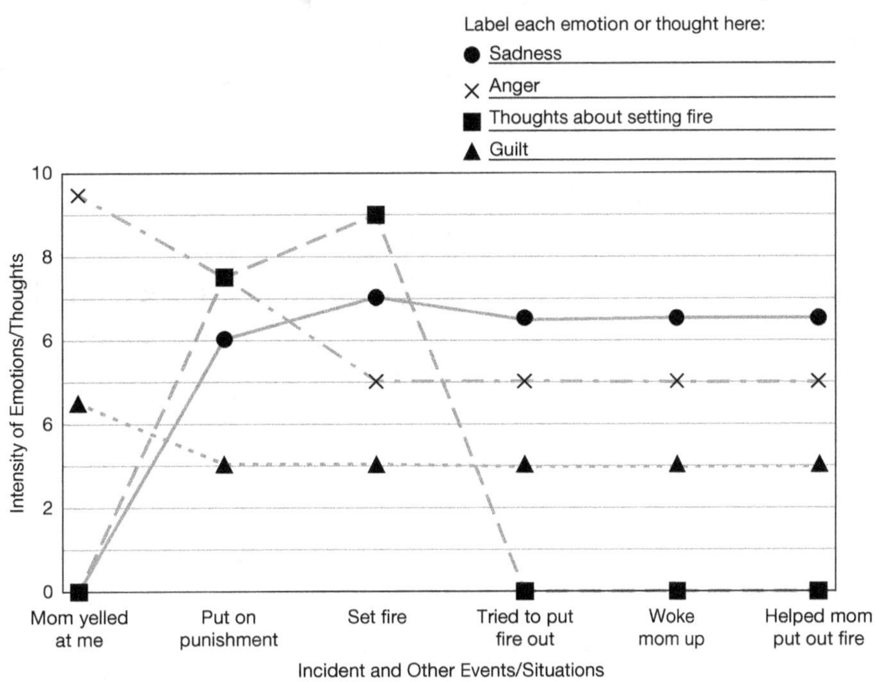

Figure 4.1

Example of completed firesetting graph.

describe the intensity of each feeling or thought along the 10-point scale, which was then initially graphed by the clinician. The clinician then sought constructive input regarding the accuracy of each graph and made revisions based on the boy's input. The final graph that you develop in this session should appear similar in format to this one.

This graphing session may begin slowly because most children initially are reluctant to discuss the affective and cognitive components of their firesetting incidents. That is one reason why it is helpful if you can provide some of the details of the incident learned from previous sources. In effect, this may facilitate a willingness to participate and also improve general recall of what has happened.

The rigorous manner in which the graph is developed should be used to lend credibility to your suggestion that the child's firesetting behavior can be understood (and possibly predicted). This procedure is also designed to support the rationale for proceeding with a skills-based approach that will train each child in specific cognitive and behavioral responses likely to facilitate self-control and more prosocial reactions to stressful interactions or circumstances.

Procedure and Probes for Developing the Graph

For this section, use the Fire Graphing Sheet, which is Practitioner Form 4.1, in appendix A.

Graph the most recent firesetting incident initially. If a more serious fire was committed before that one, then this additional fire should be graphed next.

Begin by asking the child to describe antecedents of the most recent fire and other forms of recent fire-related involvement. This information is then integrated with existing reports or measures of the child's incident. It is especially important to phrase questions in a manner that highlights or acknowledges the child's personal responsibility for the fire and to learn about the child's level of remorse as an indication of motivation for refraining from subsequent firesetting activity.

Introduce the firesetting graph by asking about a specific firesetting incident. You may begin the conversation by saying something like:

Let's talk briefly about the fire that I heard a little bit about (i.e., nature of the fire) so that we can both know what went on and what happened. I find it helpful to try to write down some of the things you say so that we can draw what happened in the fire, on paper . . . OK, so I understand that you may have been (were) involved in a fire [then give a brief description with a few details, like what was burned or set on fire and roughly when it happened].

To solicit more details about the firesetting incident, ask more probing questions like the following:

- *Can you tell me what happened?*
- *What was happening to you before the fire—what were you doing?*
- *What were you involved with? What was going on at the time?*
- *What were you feeling (how did you feel) before the fire?*
- *Did you feel anything else then? How else were you feeling?*
- *What were you feeling just after you set the fire?*
- *Did you have any other feelings after the fire?*
- *What happened after you set the fire?*
- *How did this incident end? What happened last?*
- *Did anyone notice the fire?*
- *How did others respond (later on) to the fire?*

Follow-Up Discussion

Once you develop the graph, engage in a follow up discussion:

- Review each detail carefully to verify correctness: (e.g., "Is this right—let's see, this line shows how bored you were, this line shows how angry you were, and this line shows how sad you were.").
- Clarify how the child felt at the time and the impact of the fire (e.g., "So you were really bored and had nothing to do, but you were only a little bit angry at your sister, right before you set the fire.").
- Ask if there are any more important details to the fire that the child can recall.
- The finished graph(s) should then be placed in the child's file.

Home Practice

Briefly review what you have talked about with the child and thank him for his effort. Explain that you will talk some with his parent(s) next (see chapter 5) and hope that you will get a chance to work more with him on making things better for him and his family.

Identify a short task to complete at home that would extend this material (e.g., review the details of the fire for accuracy).

Transition to Working With the Parent

Praise the child for working hard and getting to this point in the session.

Mention that the child can have some time to review the material/assignment, while you work with the caregiver ("It is now your dad's turn to meet with me.").

Parent Preparation for Treatment: Goals, Targets, and Initial Plans

MATERIALS

- "Children and Fire Program—Treatment Plan and Goals" (Parent Form 5.1)
- "How to Relax in a Few Minutes" (Parent Form 5.2)
- "Behaviors Associated with Children Who Play With or Set Fires and Some Possible Consequences for Managing These Behaviors (Parent Form 5.3)
- "Worksheet for Describing Possible Target Behaviors and Consequences to Manage Them" (Parent Form 5.4)
- Pencils, pens, and blank paper

OBJECTIVES

- Introduce the rationale for your work with the child and family, making specific reference to reducing the risk of firesetting.
- Discuss the parent's treatment goals and complete a treatment plan. Provide information about the treatment program, with special regard to reducing the risk of firesetting.
- Solicit parent perception of the child's characteristics and family circumstances.
- Assess current house rules and adult supervision of child.
- Discuss child, parent, and situational characteristics in relation to firesetting.

- Discuss parental reactions to stress and teach relaxation exercise.
- Encourage parental consideration of a home program.

Introduce a Rationale for the Purpose of the Treatment

Introduce treatment to the parent by explaining ways you will reduce the risk of firesetting. You can use the following script:

I was talking last session with your child about things in general, as well as her involvement with fire. In this session, we'll discuss some information about ways we can try to teach her some new things and to encourage her to behave in ways that you support, which will hopefully help her avoid using fire. I just started to discuss some ways to do this with your child. I would like to go over some of this information with you and to better understand how things are going with your child today, but first I need to make sure I understand your treatment goals and then we can complete a treatment plan.

Introduce the Treatment Plan to Address the Child's Firesetting

After completing the first child session and first parent session, offer the parent(s) a summary of some of your thoughts about how you would work with them. As noted in the summary of the case in chapter 3, a primary target in that case involved improving fire safety education or skills. But, in many cases, this is just one of the treatment goals to be achieved. Some common additional targets for the child include helping the child recognize emotional or cognitive contributors of using fire to address an interpersonal motive, and to improve the child's assertiveness skills. Additional targets for the caregiver include promoting access to safe activities and improving the caregiver's appropriate use of consequences and ability to maintain self-control.

Review the Treatment Plan

Parent Form 5.1 (Children and Fire Program—Treatment Plan and Goals), located in appendix C at the end of this Guide, presents a version of a sample treatment plan for use with the family. The plan has examples of typical content and goals for illustration. Review the content and then discuss the parent's goals and preference. Then, based on this feedback, revise the material to fit the needs of the family.

There are several things we can do here to try to keep children from getting involved with more fires. Some of them are things we can work on with your child. Some are things we ask parents to do. After talking with you and your child, I think it would be helpful to work with your child on learning:

- *How to identify and appropriately express feelings*
- *Strategies for anger control*
- *Different ways to think about problem situations that can lead to healthy decisions*
- *Problem-solving strategies to avoid potential fire-related situations*
- *How to be assertive rather than aggressive or passive*

I think it would be helpful to ***work with you*** *to:*

- *Develop a plan to discourage your child's contact with fire and increase your child's involvement in more safe and appropriate social activities*
- *Increase adult monitoring and supervision of your child's whereabouts and activities*
- *Learn some strategies that help you feel more confident and effective in managing your child's behavior*

In other words, I would need to meet sometimes with your child, and other times with you. Some of our sessions may be together. How does this sound to you? Do you think he needs to learn or do something else?"

Discuss Child and Family Strengths

Identify child and family strengths that might support family functioning:

Now, let's talk about your family and some of its strengths and some of its stressors. Please tell me about any important family strengths, resources, or other supports (e.g., strong cultural traditions, close family network, religion/faith financial resources). Are there any special strengths in other family members?

Ok, let's now talk about (child's) activities and interests. How does (child) spend his time? What does your child like to do for fun? How often does he get to do these things? Are there some things that he really gets excited about or will work hard to be able to do?

Next, ask how the child does in terms of:

- Following directions at home and school
- Keeping his temper when things don't go his way
- Getting along with peers (e.g., does he have close friends, does he play well with others?)
- Sticking up for himself in a good way

Learn about the Parent's Use of Monitoring, Supervision, and Discipline

At some point soon we will be reviewing useful ways or steps for helping to manage (child's) behavior to keep him safe, healthy, and happy. But before we do that, I'd like to talk to you about what kinds of things you do to monitor what (child) is doing. I know we talked about this the last time we met, but I'd like to hear how this has worked out since then.

Ask the parent to discuss these issues:

- *Who usually watches your child after school and on the weekends? Are there some times when she is home alone or with just other kids? How often does she get to go around the neighborhood by herself (or with just another kid)? Are there particular times or situations when it is hard to find an adult to keep a close eye on your child?*
- *How much time have you had direct contact with your child/adolescent over the past week or two: (1) at meals; (2) after school; (3) on weekends, and (4) in the evenings, like after dinner and before bedtime?*
- *How important would you say it is for you to know where your child/adolescent is and what he is doing at these and other times?*
- *How do you try to keep up with knowing where your child is?*
- *For those times when you weren't around your child/adolescent outside of school, was someone else around and keeping track of him?*
- *Who was it, and how does your child/adolescent respond to that person?*
- *Is that person able to enforce your rules and to get your child/adolescent to do what he needs to be doing? How much does this person pay attention to your child/adolescent and do things with him?*
- *What are the main rules you have for your child at home? How often does she follow the rules? What happens if she breaks a rule? Does she get any special feedback if she follows the rules?*

Identify Important Child and Family Stressors

Try to understand how child and family stressors might contribute to family difficulties. You can use the following script as an example:

How often would you say that it is easy and enjoyable for you to be around your child?

How often would you say that it is hard for you to be around your child?

I am also wondering about any specific individual or family difficulties that might affect you or your family. Do other things cause problems in your family, such as financial concerns, work hours, conflicts between family members, recent losses, or any other things like that?

And, have you thought about some of the behaviors or activities that (child) does that cause stress for you or your family? How do you see your child's behavior or personality affecting you or others at home?

Discuss How Child and Parent Characteristics May Be Related to Firesetting

Discuss ways that child and parent characteristics may contribute to the child's firesetting behaviors. You may use the following script:

Now, let's talk more about the fire (incident). Why do you think your child was involved in the fire incident? Do you see his involvement with fire as being related to any of the kinds of things we have talked about so far (child behavior, family burdens, adult supervision)? Could these things have influenced his involvement with fire? Which ones?

Discuss parent and family reactions to the firesetting incident, especially in terms of angry or hostile (provocative) responses to the child:

Parents and family members often have different reactions to a fire—sometimes they feel really stressed, scared, sad, confused, and/or angry. Did anything happen in your house after the fire? How have things changed?

Practitioner Note

It is important to support the parent's reasonable (understandable) reactions to the fire—be empathic (e.g., "I could certainly understand someone getting that worried about whether there would be another fire, and maybe one that was even more dangerous").

Discuss How the Parent Reacts to Stress

Discuss ways that the parent may react to stressors from the child. Begin this discussion by saying the following:

I think I have an idea as to some of the things that your child does that may be pretty difficult to deal with as a parent, and also some of the challenges for your family. I'd also like to learn a little more about what you think and feel in stressful situations, because this affects the way you might respond to that situation.

Ask what happens when the parent feels stressed in terms of:

- *Physical reactions (how your body reacts—those are signs, like, when you(r): sweat, heart races, stomach hurts or feels funny, face flushes.*
- *Thoughts you have (how your mind thinks/reacts)—like when you: have thoughts that seem a little disorganized, think negatively about yourself, think negatively about your child. You may find yourself having the same troublesome or distressing thought again and again.*

Ask what the parent is like when she is stressed. Some possibilities include:

What do you think you are like as a parent when you're stressed? For example, are you:

- *Getting angry or anxious easier*
- *Acting more emotional (volatile)*
- *Acting before you think things through*
- *Avoiding others or wanting to be alone*

Ask about any personal cues for tension/anxiety or anger/agitation.

From what you've said, it sounds like there are some times when the situation seems to get worse, depending upon how you respond initially. Sometimes, it is helpful to notice our own internal reactions to the situation and to use that information to change our behavior for the better. This might avoid making the situation worse, by hitting, humiliating, or getting more out of control than we want to be. In fact, one thing I would like to do today is to go over an easy strategy you can start using now to give your body a break from stress and tension, especially as a parent. One thing to do if you begin to feel really upset and tense is to try to keep calm; some people do this by trying to relax. This is one of the shortest and easiest things to do, just to help you collect yourself.

Teach the Parent How to Relax

Give the parent a copy of Parent Form 5.2 (How to Relax in a Few Minutes), which can be found in appendix C. Review this material together.

Here is an exercise we use to help people relax. You can do it in your home or anywhere else. We'll try one exercise to show you how to produce this feeling of being more relaxed.

Once reviewed and practiced, ask the parent for any impressions and address any questions. Then, ask the parent to practice the relaxation script at home at a time that is convenient, possibly once per day.

Review any reported negative thoughts/attributions about the child and try to find alternative ones to consider. If the parent has just acknowledged any specific thoughts (or beliefs, attributions, or explanations) that are distressing or troublesome, try to learn more about the nature, timing, frequency, and impact of those thoughts. If possible, try to identify the type of thought/attribution and learn when and how it arises, as well as what leads the parent to think that way ("what evidence suggests that your view is accurate?").

Try to discuss whether any alternative thought might be more accurate or helpful in allowing the parent to be effective in handling situations involving this child ("How else could you view this behavior?; what else could you think at that time?'). If one is identified, discuss how this thought/view might help the parent cope more effectively when the child's behavior is a challenge. It's ok if the parent isn't really sure if that alternative thought is better; you just want the parent to be aware of it and to possibly consider it as an option.

Introduce Some Common Behaviors Associated With Firesetting and Possible Strategies for Managing Them

Give the parent a copy of Parent Form 5.3 (Behaviors Associated with Children Who Play With or Set Fires and Some Possible Consequences for Managing These Behaviors), which can be found in appendix C. Review this material together. Discuss some of the possible consequences that might be used to target these behaviors. Explain any terms or concepts that the parent may not understand.

This handout lists some behaviors that are common with children who misuse fire and outlines some appropriate consequences for these behaviors. We will revisit this handout later during the treatment process.

Discuss Potential Consequences to Target these Behaviors

Ask the parent about her preferences for how she manages the child's behaviors and what consequences seem to make the most sense here. Then, with input from the parent, consider adding one problem behavior and one consequence to the "Worksheet for Describing Possible Target Behaviors and Consequences to Manage Them" (Parent Form, 5.4 in appendix C), which has been left blank for this purpose. Try to set one simple contingency in place to help manage the child's fire-related behavior. For example, it may be useful to suggest that the parent target item #5 from Parent Form 5.3 (i.e., child sets a fire that burns property or people—e.g., toys, household items, other objects, clothes, skin) with a consequence that seems fitting for the age of the child and severity of the behavior (e.g., lost privilege, contact authorities). The consequence just needs to be a reasonable one for now. In later sessions, we will review these and other parenting strategies in more detail.

Of course, note any other reasonable options that the parent suggests and consider whether to include them now or in a later session based on upcoming content. This brief plan provides at least an initial suggestion for helping the parent control the child's fire-related behaviors.

Home Practice

Ask the parent to practice the relaxation script once per day, using Parent Form 5.2.

In addition, ask the parent to review Parent Form 5.4 to help identify some additional consequences for managing the child's behavior.

Teaching Fire Safety Education: Basic Awareness Concepts

MATERIALS

- "Fire Triangle" (Child Form 6.1)
- Create a "List of Local Fire Facts" to share in session
- Pencils, pens, and blank paper

OBJECTIVES

- Help the child understand how fires start.
- Help the child to understand the impact of fire and smoke.
- Understand this child's motives for the fire.
- Understand general child motives for fires.

Session Orientation

Practitioner Note

Consider whether to include the parent and child together in this session. In many cases, having them together provides safety messages to both participants, and allows the parent to help personalize some of the terms and tasks used with the child. In our prior work, parents have found it valuable to receive this information with their children.

This educational material is primarily designed to teach the child about the nature, functions, hazards or dangers, and control or prevention of fire. Much of this information was included in a fire safety education protocol that was initially implemented in a four-session group format (Kolko et al., 1991) and later expanded for use in eight individual sessions in a larger outcome study (Kolko, 2001). It is possible that the child might already have knowledge of some of the materials. Consequently, it is important to guide the review of this material on the basis of what the child already seems to know. Also, the content is designed to go beyond simply telling the child a rule that should be followed. Instead, you should try to help the child understand the logic behind the rule or behavior to increase the likelihood of adherence to the material. Remember that when presenting fire information to children, the task may at times appear to be a boring review of facts. Indeed, it may be difficult at times to make this material interesting and meaningful. Of course, the objective is to help the child understand these facts and attend to them in the natural environment. *Remember: fire safety information is survival information.* The number of sessions needed to complete the intervention may vary from child to child. When conducting the sessions, pay attention to how much information is being presented at one time because too much information may be overwhelming, and at that point the session has gone too long. The order in which you present the information and the depth of focus on different points can vary based on your judgment and observations.

Fire Triangle—How Fires Start

At this point, introduce the concept of the fire triangle, which is an important first step in educating (or reeducating) the child and parent(s) of what it is that causes fires. The fire triangle is the interaction of the three elements that must be present in order to create a fire: oxygen, fuel, and heat. It is pictured in Child Form 6.1 located in appendix B at the end of this Guide, and can be reviewed with the child.

When presenting the fire triangle to the child and parent(s) it will be helpful to go into details about how each element contributes to the creation of a fire, using examples such as:

Oxygen is nearly everywhere. We breathe it every day. Almost any material can be a fuel, such as wood, paper, furniture, clothes, plastics, and flammable liquids. Fire starts when something hot is added to this mix, like: lighters, matches, cigarettes, or electricity.

Practitioner Note

Additional resources that could be used to teach this subject are listed here. www.smokeybear.com; www.nfpa.org/

Demonstrate the Role of Oxygen

This section is optional (a) if the child already seems to know about the role of oxygen, or (b) if there is a reason to worry that the child will get too "stimulated" by this task. If either of these "caveats" is true, skip this material and go on to the next header (Impact of Smoke).

Here is a way to demonstrate how important these three things are to a fire. I'm going to show you what happens when you add or take away air (oxygen) when there is a small flame.

Materials

- Glass pan & lid
- Small candle & holder
- Matches

Procedure

- Place candle in pan and light candle
- Begin to place lid on pan; the fire should begin to go out
- Remove the lid; the fire should get bigger
- *A fire will stop if you remove the air.*

What Can Smoke and Fire Do to People? Who Is Trained to Help?

Impact of Smoke

Any fire can cause damage or hurt someone. Smoke is also one of the main ways that a fire causes problems. It is important for the child to be aware of the dangers of not only the fire hurting people, but also the smoke. Some ways for reinforcing the idea that smoke is dangerous and poisonous to inhale include the following ("so, now we'll talk about what happens where there's a fire."):

- *Smoke prevents oxygen from reaching our brain.*
- *Smoke causes dizziness and respiratory pain.*
- *Smoke numbs the throat and can make breathing difficult.*

- *Smoke makes it difficult to see and may damage our eyes.*
- *Smoke irritates the respiratory system (how we breathe) and causes coughing and sneezing.*

Impact of Fire

It is also important to understand how fire itself is dangerous.

Fire uses up oxygen, making it difficult to breathe: As less oxygen is available, we become confused, and we have problems moving our muscles. For this reason, people often do strange (unusual) things during a fire, or have difficulty doing simple things, like opening a door. If too little oxygen is available, people can die or need to go to the hospital for special treatment.

It also heats the air to dangerous temperatures. This heat often causes people to lose consciousness or die, possibly within minutes.

So, there are different ways to get hurt by smoke or fire, such as: asphyxiation, super-heated gases, smoke, toxic products, and flames. These, and other causes, could result in burns or scalds.

Burns

There are three types of burns, which vary in how serious they are.

A first-degree burn just affects the outside layers of your skin. This type of injury isn't very serious, but it may be painful. The skin may appear to be red, but it usually heals in several days without leaving a scar. First aid for a first-degree burn involves cooling the area with running water and then putting a bandage on it.

A second-degree burn usually involves blisters and a larger area with a deeper red color. These types of burns may cause some scarring. First aid for a second degree burn involves keeping that part of the skin in cool water for a few minutes as soon as possible. It would also be important to remove any clothing covering that area and then get medical help. A medical provider may suggest a cream or lotion and a dressing that does not stick to skin.

The most severe burn is the third degree burn. It involves all layers of skin, so it could affect fat, muscle, and bone as well. These types of burns require medical attention because they do not heal on their own. They may require an operation where a doctor has to attach new skin to the area and some other kinds of special medical treatments.

Scalds

A scald is a burn that happens when your skin touches a hot or boiling liquid, like hot water. Scalds usually occur from spilled hot pots, cups of tea, coffee, soups, and other hot liquids. The recommended temperature for keeping hot tap water safe is 120 degrees F. so that it does not lead to scalds. If water is really hot, like 160 degrees, it can cause a serious, even life-threatening, burn injury.

Action Plan for Handling Minor Burns

The child should have an action plan if he should develop a minor burn. The first step to developing this knowledge is having him recall how fire can start and what kinds of things that cause burns (such as candles or stoves). Simple burns are usually just red, while more serious burns may blister. Simple burns can be treated with contact with cool water for 2 to 3 minutes. More serious burns warrant a child to tell a grown-up. If there is no grown-up present, and the burn is black or white or bleeding, the child should call an ambulance (911).

Remind the child about some ways to prevent burns:

1. Do not play with any lighters or matches.
2. Play away from a space heater, a fireplace or grill, or any area where it is hot, like a stove or cooktop.
3. Avoid touching any water or other liquids that could be hot.

For Younger Children: Identify a Picture of a Firefighter and Show It

To make the child more aware of what firefighters are and what equipment they use to keep safe and avoid injuries, it is helpful to perform an activity that helps the young child learn more about who firefighters are, what they do, where they might be spotted, and what kinds of equipment they use to fight fires. Having pictures present while doing this activity will help the child to visualize what type of equipment the firefighters use, and the types of risks with which they come in contact.

Discuss how Firefighters Have Special Training and Equipment to Fight Fires

It is important to help the child understand that fire is meant to be managed by professionals who have special training to understand how to put out a fire safely and effectively. This requires lots of preparation, practice, and coordination. Becoming a firefighter can take several years of attending special classes and completing special exercises. Part of this training teaches them how to use special equipment to keep themselves safe, to avoid injuries (like burns or scalds), and to battle a fire. They also learn how to save a person from a burning building and to take care of any people who have been burned or have trouble breathing due to the smoke. Sometimes, it may take firefighters many hours to get a fire under control. And, even when they do, the fire might have caused a lot of damage and possibly caused injuries to those who were inside or to the firefighters themselves, who were trying to put it out.

Probing questions to ask the child include:

- *Have you ever seen a real fire yourself?*
- *What was it like?*
- *Did you notice what the firefighters were wearing and what they were doing?*
- *What did you think about all of this?*

Try to offer a summary that puts several of these concepts/lessons in perspective for that child. Then, ask for and address any questions.

Review and Discuss Details of Local/Regional Fires

Parents and children may not be aware of the reality of the fire problem. Presenting facts about local fires, especially the ones started by children, will hit the message home. Parents and children may have an understanding of the dangers of fire, but they may feel distant from the problem if they haven't seen many structural fires. Before having this session, you will need to contact a local fire department or the bureau of fire to gain some facts on fires set by children or adolescents over the past year(s) or any related information to bring attention to the scope and severity of these incidents at the community level. For example, try to learn how many fires were set by youth in the past year in the general region as well as the specific town or city in which the child/adolescent lives. If possible, try to get some details about a recent youth-set fire (e.g., what was burned, how difficult it was to put it out, any injuries, extent and cost of the damages, other consequences of the fire). A brief review of this incident may be sufficient to make the point that a fire can be both damaging and dangerous.

Here's a sample of what you might discuss with the child:

Did you know that in the general area where you live, there were 23 fires that may have been started by a child or teenager? And, some of those fires caused more than $50,000 in damages and six people to get hurt. So, some of those fires were really serious. Did you ever notice these or any other fires nearby?

Explain each individual fire fact to the parent and child.

I'd like to tell you a little about the problem of firesetting, and how many fires there are in this country and what they actually do. Not many people understand how much of a problem this is. For example, you probably would be surprised to learn that... [Mention the first fact].

Review some details about the fires and discuss the child's reactions.

Take time to discuss some key details of any local fires and then solicit the child's impressions or comments. This may be done simply by making some brief comment and looking at the child to see if he has a reaction, or you can say something like:

So, I found some information about other fires that have been set in your area (neighborhood, town, city, etc.). What do you think about that (mention each fact)?

Try to paraphrase each fact in a style that seems to fit the child's level of sophistication, facilitates the child's engagement, and does not scare or traumatize the child.

Review the Reasons Why Children Start Fires

There is no single or correct answer for why a child might start fires. In many cases it is a combination of several reasons. One of the most common reasons mentioned by children is curiosity. Curiosity is especially common in children ages 2 to 7. Curiosity stems from a child's desire to experiment and explore her environment. Children want to know how fire feels, looks, how hot it gets, how it burns, its color, movement, etc. A child's curiosity about fire is an example of how the motives for causing a fire aren't necessarily to cause harm or damage. It is possible for a child to accidentally cause a fire (she may have not been aware of the elements that make up the Fire Triangle).

It is also possible that the motives were intentional. Firesetting might be a child's means for expressing herself. This type of behavior is common in children who desire attention from peers or adults. If the child is unable to achieve peer acceptance or adult attention by positive means, she may turn to firesetting behaviors. Fire may also act as a form of communication of anger or retaliation. Other examples of reasons for starting a fire include, but aren't limited to: imitation of peers or adults, boredom, little supervision, pressure from friends and peers, and reacting to something bad that has happened.

When asking the child (and then the parent) about the reasons for the fire, do not discount any potential motives and try to explore each one. Having a full understanding of all of the motives will help in the creation of the most personalized plan for treatment.

Review Child's Recent Fire Incident

Review the Investigator's Report

Reviewing the report helps you understand what happened and the context (i.e., to what extent it was serious, perpetrated by the child, and viewed as an accident). Before starting, remind the child and parent about confidentiality of all information. (Refer back to chapter 3 in this Guide, if needed.)

Have the two of you had a chance to talk about the incident? [Just to see if they agree on the fire and if they have any reactions before you get started]. As I understand it from this report (child), it appears that you were involved in a fire that occurred (name location and approximate date)... Is that how you recall it?

Discuss some of the individual characteristics of the fire, some of which might have already been known (e.g., time of fire, location, cause, what was used, severity). Write down any new details on the investigator's report. Use this information to clarify what happened and when ("just the facts"). The point here is not to document each and every detail, but to show that we have some (i.e., a limited) understanding of what happened. The discussion should clearly convey the spirit of an information-gathering session, not an inquisition in search of a confession.

OK (or, I see)... now, let me try to understand what else happened; it says that you (child) were (location)... Do you remember anything else about that place?... how did it start?... what did you use?... how serious was it?... what did you think when you saw the flame?... what did you do then?

As long as the child and parent are interacting well and there do not appear to be any adverse effects of having them together for this conversation, continue the session as a joint meeting.

Discuss Other Aspects of the Fire

Discuss any other aspects of the fire, such as the child's motive (why) and the situational context (what had happened before the incident).

Are there any other details that would be helpful for us to talk about?... Is there anything else about the fire that I should know or that you want me to know?

So, what were you thinking about when this happened?... Why do you think this happened?... I see, all of the things that were burned were..., anything else?

Summarize the Incident

Once all of the details of the incident have been documented and clarified, provide a verbal summary of the incident (that includes its relationship to the child's motivation/precipitants).

Home Practice

Ask the child and parent to learn about a fire that occurred in their community and to think about at least one aspect of the fire that they found important or meaningful. They can do this by searching the internet or contacting the local fire department.

Teaching Fire Safety Education: Strategies to Support Safety

MATERIALS

- Some materials noted in this chapter are available online and may be downloaded for review here: http://www.oup.com/us/ttw.
- Pencils, pens, and blank paper

OBJECTIVES

- Provide information about the dangers of fires and ways to avoid it.
- Teach skills to help the child reduce exposure to fire and prevent injuries or damages by responding effectively to it.
- Provide the family with materials and activities that can support their fire-safe behavior.
- Review fire project home practice assignment and summarize any "take-aways."

Discuss General Resources to Enhance Awareness of Home and Fire Safety

Practitioner Note

This material can be appropriate for both parent and child. Although some concepts may be more relevant to the parent (e.g., how to lower risk from smoking), it may still be helpful for the child to understand them.

Awareness of the possibility of having or creating a fire problem may be the first step in helping to solve or address it. Giving the child and parents a thorough understanding of some safety precautions to consider and details of a fire can help prevent any future incidents.

The U.S. Fire Administration will occasionally release a *Sesame Street*-themed fire safety guide with activities, pictures, and songs. You may find this to be a great resource, especially for younger children. The variety of resources (available in both English and Spanish) can be found at the following link: http://www.usfa.fema.gov/prevention/outreach/ss_safety_program.html.

The National Fire Protection Association (NFPA) also provides free and customizable Safety Tip Sheets, many of which are available in Spanish and other languages. These can be found at http://www.nfpa.org/safety-information/safety-tip-sheets.

The NFPA also sponsors an interactive website for children and parents with the association's official mascot, Sparky the Fire Dog. The site provides a number of educational activities, games, and videos at http://www.sparky.org/.

In addition to resources detailed in this chapter, The Arson Prevention Program for Children (TAPP-C) Manual is a great resource that includes activities, intervention plans, and details of sessions. More information about TAPP-C and how to be trained in it can be found at http://tapp-c.com/training.html.

Consider what resource would be most appropriate for this parent and child to review. Suggest that the parent review any relevant material with the child.

Reducing Other Risks of Fire at Home

To extend this discussion to other fire safety education topics, you may want to review different at-home risks and safety tips on topics such as the dangers of smoking, importance of smoke detectors, avoidance of match or lighters, safe and proper match use, and the removal of flammable materials.

Effective instruction about the risks of causing a fire from smoking is most effective if there is a smoker in the household. When appropriate, helpful safety messages to teach the parent are as follows:

- Never smoke in bed.
- Always use a big and sturdy ashtray.
- Always empty the ashtray in the toilet rather than a trashcan.

Smoke detectors are about the most effective, low-cost fire safety device a person can buy. They warn about a fire before dangerous gases make it hard for people to see, think, and move. They also wake people up in the night so that everyone has an opportunity to safely escape. Important tips to give the parent are:

- Make sure they buy a "labeled" certified and tested smoke detector.
- Make sure they are installed properly with at least one smoke detector located inside or just outside of each bedroom.
- The local fire department can sometimes install them if the parent does not know how to properly do so.
- Make sure they are maintained properly, with batteries replaced every 6 months.

Fire is a Tool, Not a Toy

Children should first be instructed that a match or lighter is a tool, not a toy. And, it is a tool to be used by adults, not children. This means that they should simply avoid trying to start a fire using a lighter, match, or other device. You can encourage the child to avoid being near fire or even touching anything that is burning or can set a fire. To highlight the principle that fire is a tool for use by older teenagers and adults, it is useful to inform or remind children of the appropriate uses for fire as a first step to educating them about how to properly use fires. Fire is appropriate for heat, cooking, and light. Teaching appropriate uses for fire by an adult or an adolescent with adult permission while also reviewing the dangers of fires and inappropriate uses for fire can be helpful in reinforcing which uses for fire are appropriate.

Some older children or teenagers may want to learn about how to use matches or an electronic flame stick lighter to help with grilling or another task where fire is involved. This is something that their parents can show them, but only if this is developmentally appropriate and an expectation for this child. The rationale for teaching a child how to light a fire is that she will understand how to use fire appropriately and safely when it is appropriate to use it. Examples of rules for safe match or lighter use include:

1. Use only with adult approval or supervision.
2. Use only for appropriate uses for fire, like cooking, heat, or light.
3. Never use for play.
4. Make sure the child asks the adult to help the child light the match.

Although these rules may be appropriate for an older child or adolescent with responsibilities related to the use of fire, all told, it may be simpler and safer to avoid teaching a young child how to use matches safely until there is a clear need to do so.

Of course, if the parent believes that this task is appropriate for an older child or teenager who may need to learn how to safety light a match, you can help parents teach their children proper match use by sharing the following instructions:

Describe these steps the next time you light a match in front of your child. Some fires start because children light matches incorrectly, burn their fingers, and drop the lit match. The steps you should teach are:

- *Open the package and take out a match.*
- *Close the package.*
- *Use the striker surface on the package.*
- *Strike away from the body.*
- *Then, after striking, hold the match (sideways) for several seconds.*
- *Blow out the flame and cool off the match before throwing it out.*

Home Fire Prevention

To support fire safety, the parent and child can be encouraged to conduct a home fire inspection. By following these tips, they can see firsthand how fires can be prevented with proper preparation. We find it helpful to review the items on the "My Fire Inspection Checklist," which is a child-friendly list that has been posted online by the National Fire Protection Association (http://sparkyschoolhouse.org/app/uploads/2015/05/SparkyChecklist.pdf).

The checklist includes a few key areas of danger, with the greatest hazard areas being smoking, heating, cooking, and storage. This form can be used as a take-home resource for the child and parent to go around their home and investigate the different items on the list and check *yes* or *no*. They can be encouraged to address as many of these concerns as they can. When they bring back the completed checklist, go over it with them, paying particular attention to the particular problem areas. Here are the items on that checklist:

- Electrical cords are in good condition (not damaged).
- Appliances and lights are plugged into separate electrical outlets.
- All smoke alarms work when tested by a grown-up.
- All escape routes are clear of clutter and easily accessible.
- Curtains and other things that can burn are away from the stovetop.
- Portable space heaters are off whenever a grown-up leaves the room and goes to sleep.
- Portable space heaters are 3 feet away from anything that can burn.
- All extension cords are used safely (not under carpets or across walking areas).
- The clothes dryer has a clean vent and filter (no lint build-up).
- The furnace has been inspected in the past year.
- The chimney has been inspected and cleaned in the past year.

Reporting a Fire

The goals of this section are (1) for the child to learn where and how to obtain adult assistance during a fire, and (2) to have the child be able to describe emergency evacuation procedures.

Identification/recognition of a fire in the area is a good starting point. Ask the child to use her senses (sight, hearing, smell, and feeling) to identify how she could figure out that there is a fire nearby. A few examples of some productive or useful answers are as follows: seeing smoke and flames, smelling burning wood and smoke, hearing a fire alarm or crackling of a fire, and feeling heat, coughing, or even choking.

After the child has a good understanding of some potential characteristics of a fire, the next task is to review a few steps to reporting a fire or other emergency. There are different steps to follow, which may vary by age or developmental level. Older children or adolescents may be able to understand and follow all seven steps noted in some recent recommendations available online (see http://www.wikihow.com/Report-a-Fire).

For other children, the following steps may be sufficient:

1. Alert residents by sounding the fire alarm.
2. Get a phone and call an emergency number for your area (911 in the United States) or ask an adult to do this for you. It is safer to use ~~a mobile or cell~~ phone from outside the burning building.
3. Report to the person who answers your call: that there is a fire, where the fire is, and the size of the fire.
4. Wait outside for the emergency personnel to arrive and let them know if anyone is trapped inside or any other important information.

It is important to make sure that both child and parent grasp some of the important "take aways" from this brief conversation, which include some "do's and don'ts" of reporting a fire. Helping the child understand that she must get help is the initial step. The most important of the "don'ts" is that the child must never go or stay in the burning building for any reason. When calling for help, she must understand she needs to do so from outside the building. The best way to report a fire is to take down the address (or, at the very least, a direction of where the building is, if it's close by), and tell a grown-up to call the fire department because there is a building on fire. Other methods for reporting a fire include the child herself calling 911 and giving her name and the address of the building, if no adult is present, or pulling the fire alarm (but only if she is certain there is a fire).

It is also important for the child to understand the irresponsibility of false alarms. If a false fire is reported the fire fighters will show up, and it will delay them from getting to a real fire, possibly risking lives in the process. In many areas, calling in a false alarm can result in legal consequences, such as being charged with a crime.

Fire Extinguishers

There are some things that we can do to help put out a small fire, and a fire extinguisher can be very effective with putting out a small fire. It may be useful at first to show a picture of a fire extinguisher (Figure 7.1) or a real one. Ask the child if he knows what that is and for what it is used. It is important that the child understand that a fire extinguisher is a tool for putting out a small fire or flame; that they vary in size, contents, function, and operation; and that we have to learn how to use them.

Figure 7.1
Fire extinguisher.

For further information, these websites may be helpful (https://www.youtube.com/watch?v=8fMxA1RhYjw; https://en.wikipedia.org/wiki/Fire_extinguisher).

For those children who can comprehend how to use a fire extinguisher, provide a brief overview of how and when to use it. The acronym "PASS" (pull pin; aim towards the flame; squeeze the handle; sweep across the fire) is an easy way for children to remember how to use a fire extinguisher in case of an emergency. If the child has any doubts or forgets how to use the fire extinguisher during the event of a fire, he should not attempt to fight the fire himself, and should quickly look for an adult instead. Additionally, fire extinguishers should only be used if others have been warned of the fire and the fire department has been called. Understanding "PASS" and the appropriate use of a fire extinguisher may be helpful in the event of a small fire.

Evacuation

A key part of fire safety education is making sure the child understands what must be done in the event of a fire in a house, apartment, or other setting where the child lives (e.g., trailer home). When teaching the child about identifying a fire in the house, it will be helpful to reinforce the material by having him answer questions such as: *"How would you know there's a fire?,"* *"If you see a fire or are in a fire, how might you feel?,"* and *"What kinds of things do people do when they're scared?"* Asking these prompting questions helps to segue into the discussion about the importance of evacuation during a fire and how to evacuate in case of a fire.

When teaching the child how to evacuate, briefly review the steps to escape safely. One document published by the Office of Homeland Security outlines several steps in the context of a common scenario involving a fire in a house (see http://www.dhses.ny.gov/ofpc/publications/documents/brochures/EDITH.pdf). This brochure shows the child how to leave the house if there's a fire by following these steps:

1. Plan your escape.
2. Meet with your family to share the plan.
3. Practice escaping during a fire drill.

A child then needs to:

- Roll out of bed
- Call or alert the family
- Crawl low in smoke
- Touch the door to feel if it's not too hot to open
- Use an emergency exit if the door is too hot
- Meet up with her family to make sure everyone is safe

When going over the steps of evacuation, have the child verbally read and practice the steps in role plays--even perhaps having him describe his own course of action. The goal of this activity is to get the child to a level of understanding whereby he can verbally evacuate himself safely from a hypothetical house fire ~~role play~~. Other sources provide additional materials to convey and practice this concept (see www.usfra.org/notes/Exit_Drills_in_the_Home) and video demonstrations (www.youtube.com/watch?v=4n1EKY_-wKI).

An additional resource can be found at: http://www.usfa.fema.gov/prevention/outreach/escape.html. It includes a home escape plan, Public Service Announcement, home safety checklists, fire escape plan grids, and other helpful materials to help teach home escape protocols.

Stop-Drop-Roll

By this point in treatment, children should have a good understanding of the dangers and causes of fires. Teaching them the proper methods of putting out a fire if their clothes or skin catch fire is an essential part of fire education. The accepted model of putting out a fire on oneself is "Stop-Drop-Roll."

When helping the child to understand the importance of these steps it will be helpful to describe why each step of the model is important.

- *Stop* because fire grows bigger with air or wind. Running around may cause the fire to grow bigger.
- *Drop* to the ground quickly and cover your face with your hands to keep heat or flames from getting to your mouth or nose.
- *Roll* on the ground over and over, from side to side "like a worm," to eventually put out the flames.

Explain to the child that, if available, it is helpful to roll oneself into a jacket, rug, or heavy blanket. After he has a good understanding of Stop-Drop-Roll, make sure the child understands that simply trying to use your hands to put out a fire is actually dangerous and should not be done because this can cause burns.

Fire Safety Project

To serve as a final wrap-up of all of the information presented, the child will create a fire safety project and present it to you and to the parents. The presentation is traditionally done during the last session, but this is up to your discretion.

The objectives of this project are:

- To teach children and their parents about basic home safety
- To establish the child's role as home safety checker
- To enhance parental encouragement/assistance

After a review of fire safety principles, clarify those aspects of fire prevention in which the child appears most interested. This interest might make for the most effective fire safety project. The guidelines of the project are as follows:

- Target an existing situation in the home or school
- Use it to correct or improve the safety of the situation
- Make sure it is developed by the child
- Ask the child to report on the project in the following session.

So, you should assign the project in this session and then review it in the next session.

Examples of Projects

- Discuss how boy scouts or girl scouts learn to use fire and what they do to keep safe around campfires.
- Discuss what parents do to keep gas cans safe.
- Learn about the training of a firefighter.
- Discuss the costs and consequences of having the fire department respond to a false or fake fire alarm.
- Discuss some ways to make sure that adults who smoke will be safe.

As noted, once the project is completed, the child will present the project to you and to her parents. It is important to ask the child to discuss what she did, how it was done, and what the outcome was. The parent(s) should be encouraged to listen to the child describe the project, but should later be asked to add additional comments that highlight the child's active and beneficial role in this task.

Review the Child's Fire-Safety Knowledge and Provide Suggestions to Apply What Has Been Learned

Once the child has learned the material covered in chapters 6 and 7, and has a better understanding of the nature, causes, and dangers of fire, it is helpful to review key aspects of this new knowledge to help set the stage for engaging in fire-safety behavior. To summarize the child's and parent's new knowledge, highlight the child's understanding of fire and the skills learned (e.g., increased competence in prevention and protection) or ask the child to do this himself (e.g., "what were the top three things you remember about fire or firesetting?"). You can also offer reminders of tips or suggestions made during the session (e.g., remind the child to be careful and fire-safe, emphasize how the child now understands more about the dangers of fire and ways in which he should behave around it).

Preparing a Babysitter or Other Caregiver

It is the responsibility of the adult of the house to familiarize the babysitter with some important details about the children and fire safety precautions. The babysitter should be shown the locations of the bedrooms where all children sleep, and all exits and the home plan. Leaving emergency telephone numbers on a pad next to the phone is essential. The parent should tell the babysitter to not smoke on the job, or use matches or lighters in the house. Parent(s) should also make sure that the babysitter understands that the children should not be left alone because unsupervised children are more likely to play with fire or matches. All of these steps to train the babysitter provide a safer environment for the child(ren) when the parent or guardian isn't present designed to ensure their safety.

Home Practice

Ask the child and parent to work together on the fire safety project and be prepared to report on it in the next session.

Basic Parenting Skills: Building Positive Parent-Child Relationships

MATERIALS

- "Do's and Don'ts for One on One Time" (Parent Form 8.1)
- "Planned Attending and Ignoring" (Parent Form 8.2)
- "Planned Attending and Ignoring Worksheet" (Parent Form 8.3)
- "Suggestions for Giving Positive Feedback and Approval to your Child" (Parent Form 8.4)
- "Rewards as Positive Reinforcement" (Parent Form 8.5)
- "How to Use Effective Instructions" (Parent Form 8.6)
- "Effective Instructions and Approval Examples" (Parent Form 8.7)
- Using "When-Then" Statements (Parent Form 8.8)
- "Possible Positive Behaviors to Encourage and Consequences" (Parent Form 8.9)
- Pencils, pens, and blank paper

OBJECTIVES

- Discuss level and use of monitoring and supervision.
- Teach and practice attending and ignoring skills.
- Teach skills for stating expectations and giving feedback.
- Discuss spending one-on-one time in a child-centered activity.
- Discuss the use of various rewards as incentives for specific behaviors.

Review of Previous Session with Parent

It is important to support the expectation that completing assignments is an important part of helping the child/adolescent and parent, and that these will be reviewed at the beginning of each session.

We talked about several things the last time we met separately. Before we get started on things for today, I wanted to check in with you about any questions or issues that have come up since that meeting. Do you have anything we need to talk about before we get started on the topics for today? Also, I asked you to start practicing a relaxation routine during the day to relax your body and mind a little. Did you try the relaxation routine? How did that go?

If the parent tried the relaxation exercise, offer praise and encouragement. If not, ask:

What got in the way of doing this?

Then, do some problem-solving and ask the parent to try again.

(For example: *It is important to take care of yourself so you can take care of your child/adolescent, and this will set a good example for your child/adolescent to complete her homework too.*)

Finally, complete the review of the previous session, and introduce the topic of today's session:

As we discussed last time, child/adolescent problem behaviors, including firesetting, often develop from some combination of child/adolescent characteristics, parent characteristics, and family stress. The fourth thing that is really important to consider in guiding children's or adolescents' behavior is the consequences that they receive. That is what we will start to focus on today.

Promoting Supervision and Discipline as Parenting Functions

Ask the parent to consider in what ways the child/adolescent's level of supervision and discipline can be increased and what positive parenting strategies can be used for support. Monitoring and supervision are especially important for times when the child/adolescent tends to get into trouble. Increasing the parent's involvement and effectiveness in setting reasonable expectations is often an important component of treatment. Also, keep in mind that punitive or coercive parenting strategies may increase children's resistance to monitoring and supervision, which, in turn, may decrease a parent's sense of effectiveness and involvement in this important component of parenting.

In this session, we will be reviewing some useful ways or steps for helping to support and manage (child's) behavior and help him to behave in ways that you find reasonable. Before we do that, though, I'd like to talk to you about any other steps that can be taken to help supervise your child. I know we talked about this earlier, but I'd like to hear if there's anything you want to try at this point.

Are there times when you think your child/adolescent could benefit from more supervision? When are those times? What are some ways that you can think of to improve supervision during these periods? And, are there also ways to simply have more time together? This is something we can further discuss.

One on One Time

In addition to increasing supervision time, it is also important to help you and your child/adolescent maintain/have a positive, warm, and loving relationship. Oftentimes, when families come to therapy, there is a great deal of negativity and frustration surrounding the child/adolescent's problem behaviors and, consequently, many of the interactions with that child/adolescent focus on these negative problem behaviors. In order to help a child/adolescent want to comply with her parents better, it can be helpful to increase the warm, positive interactions between the parent and child. The more positive the child feels about her relationship with you, the more she may be willing to work hard to comply with your rules and requests!

For Younger Children

One way to increase the amount of positive interactions is to have a special "one on one" time with (child). This one on one time can be very short (5–10 minutes), but it should be done every day. During one on one time, we recommend that you spend this time playing with (child) and using a variety of positive interactions skills, while avoiding some other things. The goal of this time is for you to spend time with (child) without directing the interaction, asking too many questions, teaching, or critiquing. Allow the child to select the activity, as long as it promotes interaction (i.e., not computer games or TV) and is unlikely to result in misbehavior (i.e., messy activities might be problematic, or games involving complex rules).

For Preadolescents and Adolescents

One way to increase the amount of positive interactions is to have a special "one on one" time with (adolescent). This one on one time can be short (10–15 minutes), and ideally it should be done daily, but at least three times per week. During one on one time, we recommend that you spend this time doing what (adolescent) wants to do, as long as it promotes interaction (i.e., not computer games or TV) and is unlikely to result in misbehavior (i.e., messy activities might be problematic, or games involving complex rules). You should try to use a variety of positive parenting skills, while avoiding some other things. The goal of this time is for you to spend time with (adolescent) without directing the interaction, asking too many questions, teaching, or critiquing.

Parent Form 8.1 (Do's and Don'ts for One on One Time) can be found in appendix C at the end of this Guide. Share this material with the parent and review together in session.

These one on one times can be a lot of fun for both parents and kids. This week, I'd like you to try to use one on one time every day, and let me know next week any questions or concerns you have after practicing.

Attending to Desired Behavior and Ignoring Minor Misbehavior

When you are with (child/adolescent), your attention may be one of the most powerful rewards and a source of motivation for (child/adolescent) to change her behavior. We call this attending to desired behaviors, or attending for short. When parents use attending they also have to use ignoring, or the attending is not effective. The way all people learn how to behave is by receiving feedback from others. This is a constant process throughout our lives.

Let's look at examples to see what I mean.

For Children

Children often learn that they can get what they want through negative behaviors, such as whining. If that doesn't work, many children will escalate into having a tantrum. For example: If a child asks to go out, his mother may say no and he may whine for a while, but his mother is trying to stay firm. He then throws himself on the floor, screams, cries and kicks. Because the mother wants to stop the tantrum before it gets even more annoying, she may say, "Ok, you can go out." The child has just gotten the message that if he throws a tantrum, he will get his way. He has also gotten his mother's attention by having a tantrum.

For Adolescents

Adolescents often learn that they can get what they want through negative behaviors, such as arguing. If that doesn't work, many adolescents will escalate into shouting, or slamming doors. For example: If an adolescent asks to go out, her mother may say no and she may argue for a while, but her mother is trying to stay firm. She may start shouting, slamming doors, or threaten to leave without permission. Because the mother wants to stop the fight or because she feels powerless to prevent the adolescent from leaving, the mother may say, "Ok, you can go out." The adolescent has just gotten the message that if she argues and threatens, she will get her way. She has also gotten her mother's attention by arguing and threatening.

That same situation could be different. The mom could just say "you can't go out because we have to have dinner," and then leave the room and ignore the child/adolescent's objections without further comment. Once the child/adolescent realizes he has no audience, the argument ends relatively soon. When the child/adolescent starts to behave appropriately, the mom starts to give him attention for more desired behavior by noticing something good in his behavior, such as saying "I really like the way that you are using a nice, polite voice."

What message did the child/adolescent get from the parent (when she walked out, turned her head away, made eye contact, used praise, etc.)? Solicit answers from the parent. This situation is important because it emphasizes the following ideas:

- Many child/adolescent behaviors are affected by the attention they receive from adults.
- Children/adolescents get feedback from whatever gets them attention.

- What gets them attention can be negative.
- If there is a certain payoff for a negative, annoying behavior, the child/adolescent is likely to use that type of behavior more often.
- Attending to behaviors we want to see happen more often helps us have more positive interactions with our child/adolescent. Over time this improves the quality of our relationship.

Teach Basic Planned Attending and Ignoring Skills

Review Parent Form 8.2 (Planned Attending and Ignoring), which can be found in appendix C. Teach the basic steps that should be followed to appropriately and effectively use attending and ignoring to influence the child's behavior.

Next, discuss concepts and identify desired behaviors that can be attended to and undesired behaviors that may be ignored. Work with the parent to complete Parent Form 8.3 (Planned Attending and Ignoring Worksheet), located in appendix C. Use information from the next section to help the parent identify ways to attend to desired behaviors and to ignore minor misbehavior.

Positive Feedback and Approval

It is important to give as much positive attention to your child/adolescent as you can because what you say (or how you say it) is sometimes as important as what you do. One of the best ways to get kids to respond to the things we ask them to do is by trying to say things that are positive. There are different ways to say the same thing, which may make it more likely that your child/adolescent will respond to you in the way you want.

Now, if children/adolescents listen to us, it's important to praise them for it. I'd like to talk to you about how to use positive statements to let your child know what you expect and like about her behavior. There are some simple suggestions for making this as effective as possible. Let's look first at ways that we can express our appreciation, provide praise, or show affection, especially with kids.

Briefly review Parent Form 8.4 (Suggestions for Giving Positive Feedback and Approval to Your Child), located in appendix C. These items were adapted from a longer list reported by Barkely (Barkley, 1979, 1987; Barkley & Benton, 2013; Barkley & Robin, 2013). Review these different suggestions, and try to identify a few in each category that the parent will use. Remind the parent that the impact of any nonverbal or verbal approval is likely to be greater if it occurs right after the child's behavior, if it mentions the child's specific behavior, and if it is 100% complimentary. Also, some of these skills are similar to those described in far more detail in a book on Parent-Child Interaction Therapy (see McNeil et al., 2011).

Using Specific Rewards

Sometimes, we need to use "special rewards" to motivate others to do things, especially when praise doesn't seem to be quite enough. Most of the time this is done to help get a new behavior started. But, once the behavior has been going for a while, the reward doesn't seem to be as necessary. Other things become more important than the reward itself.

Parents may wonder if using rewards is a form of bribery. It is not, because bribery is a word used to describe a situation in which someone tries to get someone else to do something inappropriate, undesireable, wrong, or illegal by offering an incentive. Here, we are talking about using special rewards to help the child/adolescent do something that would be appropriate and actually good for him to do—that is, something that is in his best interest. In this context, it seems more like appropriate support or encouragement.

Types of Rewards

There are different types of rewards:

- *Social*—praise (saying something nice about the child)
- *Physical*—hugs, pats on the back, smiles, thumbs up
- *Material*—tangible items (giving the child a desired snack or toy)
- *Activity/Privilege*—allowing the child to engage in fun activities

One easy way to help motivate kids to do things is to set up a simple agreement (some kind of contract) in which (child) can get access to things she wants (or likes) after doing something else she is supposed to do. Give the parent the Parent Form 8.5 (Rewards as Positive Reinforcement handout) found in appendix C. This includes a description of a reward menu. For more information about the reward menu, see Bloomquist (2006).

Using Effective Instructions

Review and discuss Parent Form 8.6 (How to Use Effective Instructions) found in appendix C.

Another strategy to help improve a child's behavior is to use Effective Instructions. We have found several "do's and don'ts" that are helpful when asking a child/adolescent to do something. Before you try to change your and your child's behavior, explain to your child that there will be some new rules or practices used at home. You may want to review how things have been managed in the past and then review, step by step, how things will be done differently for the time being. This will hopefully prepare your child for the use of these new rules/practices.

Once this review is complete, ask the parent to complete Parent Form 8.7 (Effective Instructions and Approval Examples) located in appendix C by choosing the best answer for each situation. Then go over the answers together, and talk about why one choice is better than the other.

Before moving on from this topic, ask the parent to *role-play* the following situations in order to practice using these skills. Set up the example and then ask the parent:

How could you respond in a way that would give (child) a positive message and get your point across for these situations?

- *Your child/adolescent got an A on a test.*
- *Your child/adolescent cleaned her room when you asked her to.*
- *Your child/adolescent got angry and didn't hit his sibling but did yell.*
- *Your child/adolescent hung up her coat after school.*
- *Your child/adolescent started on his homework without prompting.*
- *You see that your child/adolescent is in a bad mood but is not yelling or being argumentative.*

Using "When-Then" Statements

Give Parent Form 8.8 (Using "When-Then" Statements) found in appendix C to the parent.

There are some general rules to follow if you want to make these special rewards really count. The "When-Then" handout discusses how to use special rewards or very rewarding activities with your child/adolescent, to encourage or reinforce less rewarding and less frequent behaviors that you want to increase or promote. Here, children can learn that by doing a less frequent behavior that you want to encourage, they can then do something else, like an activity they find even more rewarding in return. That is, when your child engages in the targeted behavior or activity you have identified, your child then gets to do something that he wants. This means that there is a second activity that only happens when the first one is done. So, what this means to the child is that WHEN you do this, THEN you get (or get to do) that. The child could earn any type of reward that you could provide.

This was once called the Premack Principle, and before that, it was called "Grandma's Rule" (e.g., *when you eat your vegetables, you can go outside to play"). And, we think it is a better way to encourage positive behavior than using "If-Then" statements, since using "if" sounds tentative and suggests that the child might not engage in the behavior* (Fleischman et al., 1983). *Let's review the form.*

Selecting Potential Target Behaviors and Consequences to Increase Them

Give some thought to the kinds of things that you want to encourage your child/adolescent to do or to do even more often. Then consider using one or more of the consequences we discussed that would tell him you support his behavior, like doing something the child already likes. You don't need a formal agreement—it can simply be mentioned ("John, when you finish your homework, you can go outside and play"). Or, you can consider using a specific reward at home. Are there some rewards that you think would be helpful in changing these behaviors? What could you use initially or regularly?

Identify the Behaviors and Consequences

Parent Form 8.9 (Possible Positive Behaviors to Encourage and Consequences) found in appendix C is an interactive activity that helps parents establish a set of goal behaviors that they want to increase in themselves or in their child. The left-hand column is where the parents can list the specific behaviors they want to see increase. Some examples include:

- Having the child give the parent any incendiary (or other unsafe or dangerous) items in the child's possession

- Having the child provide the parent with a verbal account/report of activities with peers
- Having the child advise his caretaker of his whereabouts for monitoring purposes (activities involving increased supervision)
- Having the child become more involved in appropriate individual/peer activities
- Having the parent supervise fire-safe activities (such as, helping with barbecue or putting out candles after dinner).

The right hand column of Parent Form 8.9 is where the parent can list some potential rewards if the child exhibits the ideal positive behavior. The goal of this activity is to help the parent(s) take a personal interest in looking for positive behaviors and properly rewarding them. Some examples of potential rewards include:

- Access to interests (e.g., earning game time)
- Tangibles (e.g., earning a gift card for music or a download)

The bottom of the form includes some items that allow the parent to identify any special bonus rewards for the child who shows exceptional gains in several positive behaviors, any special stipulations or conditions that need to be met or followed in this program, as well as any specific responsibilities or activities that the parent has agreed to complete.

Home Practice

Briefly remind the parent about three home practice assignments, paying attention to which one may be the most appropriate to complete:

- Ask the parent to do one on one time, using planned ignoring and attending sills.
- Ask the parent to try to use praise, rewards, and effective instructions.
- Ask the parent to begin to talk to the child/adolescent about rewards and privileges he would like to earn, and what he might do to earn these.

Affect Regulation and Cognitive Control

MATERIALS

- "All About Feelings" (Child Form 9.1)
- "How Do Others Feel?" (Child Form 9.2)
- "How to Relax" (Child Form 9.3)
- "Different Thoughts—Same Situation" (Child Form 9.4)
- Fire Incident Graphing Sheet: One blank form and one completed form from the prior fire graphing session
- Pencils, pens, and blank paper

OBJECTIVES

- Generate a list of emotions.
- Name clues to determine how someone feels.
- Make connection between events (triggers), thoughts, and feelings.
- Demonstrate the nature of automatic thoughts.
- Show how specific thoughts are associated with different emotions.
- Show how different emotions (negative/positive) lead to different actions.

Review Home Practice Assignment from Earlier Session with Child/Adolescent

Ask the child/adolescent if she has completed the home practice assignment, which was to tell her parent about any three things she learned about fire in that session, and why she thought that information was important (or what it means to her).

Identifying and Recognizing Our Feelings (Emotions)

Introduce the purpose of understanding feelings and their role in problem-solving:

Today we'll talk about how you feel about different things that matter to you. And, we will talk about how our feelings can affect how we handle or solve problems. Usually, the way someone feels and thinks can affect how he solves problems. So, we may talk a little about how you usually feel about different things that have happened to you recently and how you think about those things as well.

Practitioner Note

Some of this material is meant for younger children or others who may have difficulties identifying or elaborating upon their feelings or thoughts. So, we have included a few exercises that are designed to help them recognize and label feelings more generally. If a child or adolescent is capable of doing this already, there is no need to apply that material. So, please be selective in picking the material you decide to administer here. If some or most of the material does not seem necessary, just skip to the next section.

Basic Information About What Feelings Are and Why They Are Important

You can use the following language to guide this section. These items are also summarized in Child Form 9.1 (All About Feelings), found in appendix B at the end of this Guide.

So, let's talk first about everyday feelings, since everybody has them. How someone feels about something can be very important. How we feel can have a big impact on how people respond to us or get along with us. Feelings are just part of normal life—we all have them. Feelings are inside us, so we don't see them. That means they aren't easy to identify. People often feel differently about the same thing. Sometimes a feeling is enjoyable, fun, or nice. But, sometimes a feeling can be unpleasant or upsetting in some way. Whatever the feeling may be, it is important to know them or to identify how we feel about things, to know why we have them, and to be able to talk about them. Can you think of any feelings you've had in the last 24 hours (last week)?

Recognition of Other's Feelings

Review Child Form 9.2 (How Do Others Feel?), which can be found in appendix B. Read the examples aloud and ask the child to fill in the blanks with the emotion he would feel in each situation. Try to get more than one response from the child for each example. Ask the child if he ever feels the same way and, if so, learn the circumstances. Review enough examples to determine if the child has this capacity.

It is also important to learn how other people feel about things because if you can figure out how people feel, then you can solve problems without creating new ones.

Using Facial Expressions to Identify Others' Feelings

For Younger Children

Find 5 to 10 pictures of children with different types of facial expressions. Colored pictures are preferable, but any will do. You can certainly make some faces and ask the child to tell you how you may be feeling to see how well she can identify feeling states. Try to pick or make facial expressions with the most basic emotions (e.g., happy/glad, sad, mad, worried/anxious) and a few more complicated or complex ones (e.g., confused, embarrassed, shocked, overwhelmed). Show each picture and ask the child to identify and describe the feeling depicted in the picture.

As a second part to this exercise, review the pictures and ask if he has had the same feeling as shown in the picture. If the child is amenable, try to solicit details as to when and why he felt this way.

Since our feelings are inside us, the only way to determine how someone is feeling inside is to look on the outside. If you can recognize feelings in others, you can also decide how you want to act with them. One way to do this is to look at peoples' faces when they are talking and then even when you're listening. We learn a lot from looking at faces—certain parts of our faces are used to express different emotions.

Here are a few probes to determine how well the child can identify "what is this feeling?" and "have you ever felt this way?:"

- How do you think the person is feeling? (from left to right: sad, angry, worried, happy)

- What do you think happened to make him or her feel that way?
- Have you ever felt this way?

If you need some additional material, show the child a few examples with your face and then have her practice showing you similar emotional expressions. You can ask the child to demonstrate some of these "for the camera." Then, just act as if you're taking several pictures of the child "making a face."

For Older Children and Adolescents

For the older child or adolescent who may need some extra help in recognizing feelings in others, you can use the same photos to ask the child to identify feelings from these facial expressions. Or if it is more developmentally appropriate, you can show the older child/adolescent some pictures of young people with different emotional expressions on their faces from a magazine, book, or other resource.

Other Ways to Learn How Someone Feels

You can also learn about someone's feelings by listening carefully to the sound of their voice. Let me show you the ways that my tone of voice can tell you how I am feeling (briefly demonstrate here using different vocal sounds: loud = angry, soft = sad).

Now, if the way I sound doesn't agree with the words I've said, it makes it tougher to figure out how I may be feeling. When this happens, you may have to ask me how I'm feeling. What are some questions we can ask someone to see how they are feeling? Here are a few as well.

Examples of questions to ask:

- *How are you?*
- *What are you thinking?*
- *You sound ______, are you?*
- *What's up?*
- *Do you want to talk?*

People May not Always Say How They Really Feel About Things

Sometimes people may not say how they feel about something because they worry that others might not respect, appreciate, or support them if they knew how they really feel. Or, they may not be sure how the person will respond. It isn't easy to share that kind of personal information with friends or family, since you never know how they might react. For example, your friend at school might laugh after seeing an older student tease another classmate, even though your friend may actually feel embarrassed for the classmate. Do you think this happens? In which situations do you think people might do this? Why do you think people might prefer to keep their feelings hidden? Have you ever reacted a certain way, but actually felt differently? Has this ever happened to you?

Identification of the Child's Recent Feelings from Pictures of Faces

Briefly Review the Completed Fire Graph (chapter 4) to Explore Precipitating Feelings/Emotions

Try to solicit several details of the fire that were discussed in an earlier session. You can remind the child/adolescent about key details, if there is minimal recall, in the hope that these stimulate other information, so have the completed version of Practitioner Form 4.1 (Fire Graphing Sheet) handy. Try to both confirm the accuracy of these details and solicit any others to help identify the feelings or thoughts the child experienced before the incident.

Then, briefly summarize what the child felt and thought before the primary (referral incident) fire, as well as what the child actually did to burn something or set the fire. Include a brief summary of any similar information learned from any other incidents of burning or firesetting.

We talked last time about some of the things that you did when you set the fire and some of the feelings and thoughts you had before you set the fire. Do you remember some of the things we talked about? Good. What do you recall? I also have the graph here and will add a few more details based on what you tell me.

If not much emotional detail has been given, consider including a few probes to learn what were *the outside triggers* (what someone did that contributed to the child's feelings) and any *inside triggers* (anything the child said to herself—e.g., self-statements or instructions) that the child reports before setting the fire. Both of these are "warning signs" that signal that the child was starting to feel badly or get upset about something. You can also ask if the child noticed certain *body cues*, like feeling tightness in the chest, a knot in the stomach, making a fist, or clenching her teeth, which signaled that she was starting to get angry or anxious.

Ask the child if she has had similar feelings at other times recently and, if so, request some details (e.g., when, who was involved, where, what happened, what she felt and thought, what she did next, any consequences, how it ended).

Use this information to identify a key feeling or emotion that the child routinely experiences that may sometimes cause problems for the child. If this emotion is related to heightened anxiety, anger, or frustration, prepare the child for affect management work.

Getting anxious or angry is our body's way of telling us that we may have a problem. When we're very angry or anxious, we may be less likely to "think carefully or safely" about what we're doing, which may cause us to do something that could hurt others or damage something. But, if we can quickly get calm or settle down, we can better manage what we do when we're feeling very angry or anxious.

For those times when you feel these emotions and these physical reactions, you can do two things to calm down—use deep breathing and muscle relaxation. For those times when you react with certain thoughts, you can learn how to challenge your thoughts and figure out if a different one is more helpful.

Teach Muscle Relaxation as a Way to Manage Distressing Emotions (When Needed)

Provide a rationale for using muscle relaxation to stay in control.

Now we are going to spend time learning and practicing new ways to help you stay in control when you get angry or anxious. You can teach your body how and when to stay calm. When people are angry or nervous, they breathe in a shallow, quick way. When people are nervous they tend to hold their breath or take short/quick breaths. So, they don't get enough oxygen into their bodies, which can result in feeling dizzy, lightheaded, and confused. Deep breathing works well to help us calm down.

Muscle relaxation works to calm our bodies down by getting rid of the tension in all parts of our bodies. When we're upset or anxious, our legs tense up because they are ready to kick or run, but we may get this way even when we are not in danger. We have to help our bodies know when to be relaxed and when to be tense. So, you'll learn how to tense up and then release different muscles in your body.

Practice Relaxation

Practice muscle relaxation by modeling the technique and then by coaching the child to use the technique, as shown on Child Form 9.3 (How to Relax), found in appendix B. First, teach the child to breathe deeply and slowly using full measured breaths. Repeat this deep breathing sequence a few times. Ask the child how he feels after this practice. Then, go through the relaxation steps and tense the body or specific muscle groups. Once reviewed and practiced, ask the child for any impressions and address any questions. Then, ask the child to practice the relaxation at home at a time that is convenient, possibly once per day.

Discuss Practicing and Using Relaxation

Discuss when (what circumstances, settings) and how (when the upset feelings are first identified) the child could use relaxation. Make sure to address any obstacles the child has to being able to breathe slowly (e.g., high level of anxiety or discomfort, difficulty sitting in a chair without fidgeting) by offering any helpful or supportive variations for completing this exercise (e.g., lie on the floor with the lights on and eyes open, use a large straw).

Introduce Automatic (Instant) Thoughts

Illustrate the Influence of Thoughts on Feelings

Sometimes when we notice having a "not-so-good" feeling, it may be helpful to think about what went through our mind just before we had that feeling. That's because what we think can influence what we feel.

Here's a way to understand how this works. Can you make yourself angry/sad? If so, please tell me how you did it. Did you think about anything special to feel that way? Sometimes this type of thought may happen more than once and it may happen pretty quickly, without really having been aware of it. This is called an "automatic thought"—it just comes to you automatically (instantly) or right away into your head. Has that ever happened to you? This seems to happen to everyone, even though we may not always know when it has happened.

People Have Different Types of Automatic Thoughts

Review the situation described in Child Form 9.4 (Different Thoughts—Same Situation), located in appendix B.

Let's take an example. Imagine waking up in the middle of the night and hearing an unexplained noise outside of your window. What might you think first? And, then, how might you feel? (Scared.) What is it that might bother you—the noise, or the thought that the noise might mean that there is a burglar out there? (Use guided discovery to get the child/adolescent to say that it was thoughts/images.) What will you do if you think it's a burglar? (e.g., call for parents, grab something heavy.) What might you think about that noise if you knew that it was New Year's Eve? (e.g., fireworks, nonfrightening cause.) What would you then do? (Go back to sleep.) So, how we think may influence how we feel about something and how we may react to it.

Another example of an automatic thought is the first thing that comes to mind when you get hurt. Say you fall off your bike; you might automatically (and instantly) think, "Ouch, that hurts."

Or, another example is when you see someone helping another person, like opening the door to help someone get into a store, and you say to yourself, "That was a very nice thing for that person to do."

Or, another example is when you see a lit candle sitting next to a backpack on a kitchen table. You might think, "That's dangerous—it could cause a fire." That's another example of an automatic thought—it's just the first thing that pops into your head. This just happens to us as people, as we get more experience with certain situations.

For children/adolescents who responded well to graphing the fire incident but now need more practice here, show and refer to the previously completed fire graph and discuss any possible automatic thoughts that may have followed each event or emotion on the graph. Then, using a new piece of graph paper, graph a more recent firesetting incident or another stressful event and discuss the relationship of thoughts to each event and emotion on the graph.

Examples of Things or Thoughts That Can Make Children Upset

Sometimes, children/adolescents have thoughts about things that bother them or thoughts that make them upset. And sometimes they think about a situation one way, but later think about the situation in a totally different way. This usually happens when people do not understand what is really going on and they are stuck with one negative thought, so they may not be able to solve their problems easily.

- *"If you think that someone is mean, you'll respond by doing what . . .; but, if you think that someone is nice, you'll respond by doing what . . ."*
- *"If you think that vegetables are bad for you, you'll never feel good about them and you'll probably never eat them."*

- *"If you think that someone is out to get you, you will probably never talk with them and you may even try to hurt or get back at them."*

Automatic Thoughts Are Sometimes Correct/Helpful and Sometimes Not

A lot of automatic thoughts develop about different people because of the way they have acted before. Like, when your mom and dad are mad, you might automatically assume that it is because of something you did. Have you ever thought this way? Later, you may find out that they were mad at someone else, or they had a bad day. But, for the time being, they seem mad and angry, making you think that it is something that you have done. You can see that maybe what you were thinking about why they were mad wasn't really the reason at all. Sometimes the automatic thoughts we have turn out to not be accurate or correct. Have you ever had a time when you misunderstood what was going on, or had an instant thought that you later realized wasn't accurate/correct? What happened? What was the thought you had first, and then what was the new thought or explanation you had? Did having the new thought help or not—and, why?

Show Figure 9.1: Automatic thoughts.

Figure 9.1
Automatic thoughts.

Here is an example of how this might work. In this picture, what we see depends on the way we are looking at it. What do you see in the first picture? Can you also see the other image? (Mention the picture the child didn't say—i.e., 2 faces looking at one another vs. a stand or table from the side). This picture helps to show how we could see the same thing very differently depending upon the one feature we notice.

Sometimes we might think we understand what is happening, but we might not really understand it. For example, we might think we know why someone did something, when there may be a totally different reason for doing it. It's important to figure out the reason why someone might have done something and not just the one reason we come up with first.

Linking Automatic Thoughts and Emotion Identification to Problem Solving

When kids think in a way that uses all of the information they have (facts, details, what they observe, what others tell them), they are starting to show good problem-solving. To think like that, you have to pay attention to how your body is responding, what feelings you are having, and what your mind is thinking. You might have to work hard to figure out what you're actually thinking. Once these things are recognized and understood, it can be easier to know when you're having automatic thoughts that might not be accurate or helpful.

Home Practice

Consider using any option that seems more useful for this child:

1. Ask child/adolescent to bring in two photos that show feelings that were very special to him and to explain both the feeling being expressed and why he chose the picture.
2. Have the child write down three situations that made him angry or anxious and explain why he felt that way.
3. Have the child practice relaxation and report back on its use and benefit.
4. Ask the child to write down a situation in which he recalls having an automatic thought and what the thought was.
5. Ask the child/adolescent to identify a situation in which he had an automatic thought that he later thought might not have been accurate or helpful. And then, he should try to figure out if there's a more accurate and helpful way to look at the situation.

Effective Problem-Solving

MATERIALS

- "A Problem is Something That Happens Between People..." (Child Form 10.1)
- "Identifying Problem Situations" (Child Form 10.2)
- "Alternate Uses Worksheet" (Child Form 10.3)
- "Solutions and Consequences Worksheet" (Child Form 10.4)
- "My Favorite Things Worksheet" (Child Form 10.5)
- "Problem-Solving to Prevent Fires Worksheet" (Child Form 10.6)
- "Problem-Solving Worksheet" (Child Form 10.7)
- "Practice at Home Worksheet" (Child Form 10.8)
- Pencils, pens, and blank paper

OBJECTIVES

- Help child to recognize problems.
- Briefly review characteristics of problems.
- Introduce the first two problem solving steps:
  - State exactly what the problem is.
  - Decide on your goal.
- Practice interpreting problem situations in many different ways.
- Introduce the last four problem solving steps:
  - Think of as many solutions as you can.

    - Teach child to think about likely consequences to judge the effectiveness of solutions.
    - Choosing the best solution.
    - Making a plan.
- Examine actual problems with which the child has difficulty.
- Practice the problem-solving steps to develop practical and safe solutions.
- Have child consider choices of good and poor solutions to a practice problem situation.
- Highlight predicted consequences of individual behaviors in leisure time activities as a way of improving the child's ability to anticipate outcomes.

Explain the Meaning of a Problem

Introduce the child to the concept of a "problem" by sharing Child Form 10.1 (A Problem is Something That…), found in appendix B at the end of this Guide. This form illustrates the 3 characteristics of a problem.

Today we're going to talk about problems and how to solve them. A problem is something that:

1. *Makes someone have unhappy or upset feelings*
2. *Happens mostly between people*
3. *Needs to be solved (we have to do or say something to make the problem go away).*

After the child understands what a problem is, you can ask him to list any types of problems and their associated feelings. This exercise will help the child become familiar with what kinds of feelings are usually related to which situations. Either you or the child can write down these examples (for most problems there should/could be more than one feeling).

Next, assign the child a distressing or "negative" emotion and have him come up with three different problems that could or would result in that emotion. You can use the following script:

Imagine that you're feeling really disappointed. What kinds of things would have happened to make you feel like this? [If the child has a problem coming up with an example, you can consider giving these examples—*Maybe your friend had to cancel a plan to do something last week? Or you didn't get the grade you expected on a math test?*]

Steps 1 and 2: Identify the Problem and Goal

Discuss Initial Problem Solving Steps

Review the definition of a problem.

So, a problem is something that happens between people, gives someone unhappy or upset feelings, and needs to be solved. Today, we're going to talk about how to solve a problem you have. You might even have a problem in mind we could solve. If so, you can tell me what it is, how it makes you feel, and what caused this problem. Although thinking about problems may be upsetting, thinking about your goal for how the problem can be solved can help.

State the first two steps:

- Step 1: Say what the problem is.
- Step 2: Pick your goal—what you want to have happen.

We must recognize that an upset feeling is happening and label that feeling to begin solving the problem. How can you tell that you're feeling upset? What nonverbal cues (body language, facial expressions, and so on) can you use to recognize feelings? The first step to problem-solving is to say what the problem is. The second step to problem-solving is to decide on a goal—what you then want to happen.

Have the child practice these two steps by asking these same questions about each scene:

Let's say that your friend borrows your jacket and says she lost it, but you see her wearing it after school. So, what is the problem, how would you feel, and what would be your goal? [Note: There are several ways the problem and goal can be identified.]

Example of identification of the features of a problem:

- Problem: Don't have jacket; friend lied.
- Feeling: Angry, disappointed.
- Goal: Get jacket back, talk to friend about what happened, come to a resolution.

Here are some examples of other problems. Tell me what the problem is, how you'd feel, and what your goal would be:

- *You are watching TV and your sister comes in and changes the channel.*
- *You lent your friend a video game and he lost it.*
- *Your mother punished you for something you didn't do.*

To ensure that the child understands problem-solving, ask her to determine whether or not a problem needs to be solved. You can use the following script:

Sometimes we might have a problem, but it is not big enough to be worth solving. For example, let's say that you have a friend over and she wants to watch a movie you really like, but you've seen it before. You might disagree about which movie is your first choice to watch, but you don't mind watching the movie again. There might be a small problem here (disagreeing about which movie to watch), but it may not be a big enough problem that it needs to be solved. We call choosing which problems to solve "picking your battles."

Ask the child if there was ever a time when she had a problem that was just not worth arguing or fussing about. What was it?

Remind Child about Role of Thoughts and Interpretations

Remember what we said earlier—that problems can be interpreted or understood in many different ways. The way you explain what's happening in a problem (how you think about it) can determine what your goal will be. You might have different thoughts about each problem, which might mean that you could have different goals; it becomes important to choose your goal carefully.

Present Child Form 10.2 (Identifying Problem Situations), found in appendix B, and ask these questions about each image:

- *Now, what is going on here?*
- *What feelings could he or she be having? Are there others?*
- *How can you tell? Can you recognize their feelings by looking at their faces and what they're doing with their bodies?*
- *Is this person having a problem? Is it worth solving?*

Steps 3 to 6: Problem Solving Solutions and Consequences

Step 3: Generate Solutions

The third step to problem solving involves thinking of as many solutions as you can. A solution is a way to solve a problem. This step is important because there are many different ways to solve a problem: the more ideas you think of to solve the problem, the better your chances are of solving the problem successfully. If one plan fails, there is always another plan available. Thinking of many different solutions prevents you from using the same solution to solve different problems.

Practitioner Note

Practice identifying different solutions with the child using this optional section.

Ask the child to come up with alternate uses to things using Child Form 10.3 (Alternate Uses Worksheet), found in appendix B. To practice flexibility in thinking, have the child think of different uses for ordinary objects. Do one example and then ask the child to list as many uses as he can for 5 to 10 objects. Uses must be specific, nonredundant, and acceptable. If necessary to boost motivation, introduce some session incentives contingent upon participation and being creative (e.g., stickers, free time). This exercise gives you an idea as to the child's intellectual ability, verbal expressiveness, and general motivation to learn the material. You may use the following script:

One way to get good at thinking about solutions is to be able to think quickly about things. One exercise we use involves trying to think of many different ways to use objects. Try to think of all the ways in which you could use a . . .

Go through the first three problem-solving steps to help solve a typical problem. Ask the child to think about a boy who was not invited to a get together that night with a few of his good friends who are going to a mutual friend's house:

- Review each step.
- What is the problem? What is the goal? What could be done to solve the problem?

- Once one solution has been suggested, ask, "Anything else he could do?" until at least a few ideas have been suggested by the child.
- Summarize the solutions and offer any feedback on the exercise.

Step 4: Figure out the Consequences of Each Solution

Introduce Step 4: Think ahead to what might happen next.

Now that you know about coming up with a lot of solutions, we're going to talk about how good those solutions are, in other words, how well each of them could solve your problem.

Using Child Form 10.4 (Solutions and Consequences Worksheet) in appendix B, ask the child to write out the solutions and consequences.

- This is the time when you need to carefully consider whether the solution is likely to be helpful in solving this problem.
- If the answer is unrealistic, probe for a more appropriate answer by asking, "What do you really think would happen?"
- If the consequence is realistic ask, "Would you want that to happen?" and "Could this happen too?"

Step 5: Make a Plan of Action and Try It

Once you have chosen a solution to try, you need to make a plan for using that solution. It is important to write out each step to the solution so that you have a plan of action for solving the problem!

Emphasize the importance of knowing the details about the required behaviors, materials/people needed, and the dates and time to implement the solution. Here are some questions to ask that address this step:

- *What needs to happen next?*
- *Once this is done, then what?*
- *When do you think this might happen?*
- *Will you need someone to help you do that?*
- *How long do you think this solution will take you?*

For practice, have the child pick a solution to one of the problems on the "Solutions and Consequences Worksheet" (Child Form 10.4), and generate a plan of action to implement the solution. Have the child write out the details of his plan on a blank sheet of paper. Encourage his efforts to employ behaviors he may not usually do (e.g., telling the truth to his friend about the lost comic book).

Step 6: Evaluate the Outcome

Introduce Step 6: Understand the positive and negative consequences of the solution.

After you've tried out your plan, you will then need to do another important step. Evaluate the outcome! Did your plan work/did you reach your goal without making more problems? If your plan worked, pat yourself on the back and remember this solution in case you have a similar problem again! If your solution didn't work, think about reasons why, and start problem-solving again to come up with a different plan. This may involve talking with others in your family who may have been involved in the problem and trying to discuss what happened and what oher solutions might be considered next.

Practice: Problem-Solving Applications

Review Recent Difficult Situations and Use Steps to Develop New Solutions

Ask the child to describe at least one problem situation that recently occurred, hopefully one that was fairly recent. The situations do not need to reflect involvement or interest in fire. It is important to guide the child through the problem-solving steps in order to show her some helpful things she can do that may solve problems easily, without creating new problems. Use the following script to learn about the child's recent problem situations:

Now that we've learned about the steps for solving a problem, let's talk about a difficult experience you may have had in the last few days, last week, or even before that time. Did anything like this happen to you? Which one do you prefer to talk about first?

If necessary, ask about details from the situation(s) to ensure that you understand their entire context (e.g., "Are you saying that you hit him first or he hit you first?"). If the fire incident graphing technique was useful for this child in previous sessions, you may want to graph this/these incident(s) to get details about the order of events and associated thoughts and feelings.

If the child has trouble generating examples of recent difficult situations, ask him about possible problem areas that you know about already, such as conflicts with siblings, parents, or teachers; or getting into trouble at home or in school. If a parent has told you about a recent "difficult situation" the child faced, you can bring this up too.

Next, ask what the child did in that situation and the reasons for his choices. Here are some discussion probes to consider:

- *How did you feel when that happened?*
- *What was your problem in that situation?*
- *What was your goal there?*
- *What did you do next?*
    - *Did you think about multiple solutions/other ways to handle the problem?*
    - *Did you think about the possible consequences of each/what would happen afterwards?*
- *What did you choose to do?*
    - *Why did you make that choice?*
    - *What was your plan?*
- *Did that make a good solution?*
    - *Why or why not?*
    - *[If not:] What else did you think you might do?*

When reviewing why the child chose a particular solution, pay close attention to the development of any alternative solutions by asking several types of questions/probes:

- *What might happen next?*
- *Would you want this to happen?*
- *Is this solution a good one?*
- *If applicable: How would this make the other person feel?*
- *What is likely to happen later on or in the long run?*
- *What else could you have done?*
- *What else could a person in this situation have done?*

Give the child some feedback as to the most likely consequence for each solution and compare different ones, guiding the child through a series of probes. Then, try to have the child consider one or two prosocial responses that he could use next time. Here are some examples:

- *Tell the kids that you don't want to play with fires any more.*
- *In that situation, it sounds as if there are a few different ways to solve it. You already (restate solution), but you could also have (suggest two alternatives). If you use this solution, you will likely (give example outcomes for each solution).*
- *Now, if you're thinking ahead to what might happen next, then you're going to ask yourself: Would I want that to happen? So, which solution would you try first?* [If child is uncertain, point out that there may be more than one good way to solve a problem.]

Have the child role-play the selected solution:

This would be a good time for you to act out this solution and to see what it is like. Now, don't forget to think ahead about what is going to happen if you really do this.

Review performance and encourage actual use of this practiced solution.

Application of Problem-Solving Steps to Leisure/Activities

Ask the child to complete Child Form 10.5 (My Favorite Things Worksheet), found in appendix B.

I have a form that I'd like you to fill out. [Offer to fill it out for the child if necessary.] *It will help us talk about the things that you like to do when you're at home or outside.*

After the child completes the handout, ask follow-up questions to clarify the kinds of activities she likes to do (or would like to do) and how often she is able to do things she likes. Explore times where the child may be bored or lonely. The goal is to apply the problem-solving steps to situations involving leisure activities to help the child identify and gain access to activities likely to promote prosocial behavior. Some examples are:

Let's try using the same steps as before, but this time let's talk about a situation where you don't have much to do or where you're trying to figure out how to spend your time. Can you think of any situations like this? Some examples might be sitting in your room without much to do or walking through the neighborhood by yourself.

Application of Problem-Solving Steps to Fire-Related Situations

Apply problem-solving steps to a situation involving fire or firesetting behavior.

Use "Problem Solving to Prevent Fires" (Child Form 10.6 in appendix B) to practice problem-solving in a situation related to firesetting. These situations should be based on the assessment findings and the child's reports about the context in which she has played with fire previously. You can use the following script:

We also need to think about how these problem-solving steps can be used in situations where you might think about or become involved with fire. Let's do this worksheet together. Can you think of any possible fire situations we can talk about? Maybe you saw matches on the kitchen table or noticed your friends playing with fire?

Home Practice

Here's a worksheet for you to take home and work on. It is a reminder of what we talked about today and it has some examples for you to think about. A lot of kids find it helpful to do this for a few minutes just to get some practice solving problems on their own. I can take a look at this next week and we can talk about some of the goals that you thought were important.

There are two options for home practice. Consider which one is more relevant to this child.

1. Give the child the "Problem Solving Worksheet" (Child Form 10.7 in appendix B) and briefly review what the child should do for each question:
2. Using the "Practice at Home Worksheet" (Child Form 10.8 in appendix B), ask the child to come prepared next time to discuss a few problems that were difficult to solve this week. Next week, you can ask the child how he handled these problems and how he might have applied the steps he learned to this situation. There can be any kind of problem. Here are some examples:
   - Having to choose between two different ways of handling something
   - Having some kind of argument or disagreement with someone
   - Having to decide about doing something that might get the child into trouble, like taking something or playing with fire

Basic Parent Management Skills: Using Consequences to Correct Problem Behavior

MATERIALS

- "Using Logical Consequences Worksheet" (Parent Form 11.1)
- "Removing Privileges Worksheet" (Parent Form 11.2)
- "Possible Problem Behaviors to Discourage and Consequences" (Parent Form 11.3)
- "Point System Program" (Parent Form 11.4)
- "Potential Challenges to Implementing a Home Management Program" (Parent Form 11.5)
- Pencils, pens, and blank paper

OBJECTIVES

- Review the use of logical consequences.
- Review ways to withhold privileges.
- Review the use of a point system at home.
- Develop a home-based contingency program.
- Review common obstacles to home programs.

Review of Previous Parent Session

Last time we talked about strategies that you can use at home with you child/adolescent to encourage positive behavior. Did you try some of these? How did they go for you?

Strategies from last caregiver session include:

- Giving attention to positive behaviors
- Ignoring minor misbehaviors
- Giving instructions in a positive, specific way
- Rewarding good behaviors
- Using "When-Then" statements
- "Planned Attending and Ignoring" (Parent Form 8.2)
- "Planned Attending and Ignoring Worksheet" (Parent Form 8.3)

Logical Consequences

Rationale

In this session we will be talking more about consequences, especially for correcting or managing behaviors we want to discourage. A consequence is anything that happens as a result of a child/adolescent's behavior. Last time we met, we talked about some of the ways you can encourage (child name) to do things using special consequences, like rewards. Of course, sometimes we need to discourage children/adolescents from doing things we don't like—things that are unwanted because they are inappropriate, manipulative, dangerous, or just plain annoying.

One way to do this is to expose children/adolescents to logical consequences. If the consequence is logical and related to the behavior in some meaningful way, it should make more sense to the child and may be more effective. A logical consequence can be effective by teaching a child/adolescent a good lesson or by helping her learn about responsibility. Obviously, the consequence that is given will depend upon what the behaviors are. Of course, we never want a child to experience a consequence that is painful, dangerous, or likely to result in getting hurt or having an injury. So, it's also important to balance the impact of the consequence with the safety of the consequence.

Review Parent Form 11.1 (Using Logical Consequences Worksheet), found in appendix C at the end of this Guide. Ask the parent to complete the form, following the directions and going step-by-step.

Let's take a look at this worksheet which discusses the use of natural and logical consequences. The Worksheet says that logical consequences can help correct irresponsible or immature behaviors. When experiencing a logical consequence, your child must deal

with the effect that is determined by you, the parent. As you select a logical consequence, the idea in general is to, "let the punishment fit the crime." It is important to make sure that the consequence is meaningful to the child/adolescent. But, you also have to make sure that your child does not experience any consequence that could be dangerous, hurtful, or harmful, or that is in any way too harsh. For example, you cannot deprive a child or adolescent of basic necessities like clothing, his/her bed, sleep, food, or water, as well as safety and security. And, you wouldn't want your child to be in pain, hurt, or injured to teach a lesson. So, we don't rely upon the natural consequences that might typically result from a dangerous behavior, like playing with fire, because the child or others could be hurt or even worse.

Removing a Reward or Privilege

Rationale and Description

A child/adolescent may learn to stop engaging in an undesirable behavior by having to lose a special privilege or simply experiencing the natural consequences of an action. Here, the child/adolescent would lose a previously earned (or soon to be earned) reward after doing something specific that you don't like. This is especially helpful if privileges can be used. The privilege to be removed should be one that the child/adolescent likes and often gets, and can be easily removed by you. Again, make sure that the consequence is meaningful, without being excessively harsh.

Here are some examples of losing or removing privileges:

- *"If you hit someone, you may lose the opportunity to visit that person again."*
- *"If you burn a toy, you lose that toy."*

Do you think these are appropriate consequences? What would be some examples of appropriate removal of privileges? What are some examples of inappropriate removal of privileges?

Give the parent the "Removing Privileges Worksheet" (Parent Form 11.2) found in appendix C, and review it together.

I also want to emphasize a few key points:

- *Make the loss brief.*
- *Clarify the specific criterion for loss and return.*
- *Be fair and quick (no extra nagging, etc.).*

Making Restitution

In general, restitution means that a person is asked to fix or replace something belonging to another person that one has damaged, lost, or stolen, or to provide some compensation after causing that person to suffer a loss or an injury. If you borrow someone's toy and break it, you can pay for or replace it. Or, if you accidentally lose someone's hat, you could also replace it. This is usually a fair and quick way to replace a lost item.

In the case of a fire set by a child, it is possible that someone lost property and/or sustained an injury, among other consequences. It may not be feasible for the child to make restitution for the injury or for something that is really expensive or that simply can't be replaced. But, it may be possible to make restitution for some or all of the lost items belonging to someone else that were burned by that child's fire. So, it may be a helpful and corrective consequence to ask the child to either pay for the lost item(s) or to give something up of either similar monetary or sentimental value. That is, if the child's fire burned his brother's favorite baseball mitt, it might be fitting to give the brother the child's own baseball mitt as a replacement.

Ask the caregiver if the victim has already asked the child to make restitution and, if so, learn the details. If not, you can ask the parent to consider some form of restitution that would compensate the victim here in some fair way.

- *Do you think your child would benefit from some type of restitution here?*
- *Could the child pay to replace (the burned item)? If so, what would be a reasonable or fair price?*
- *If the item can't be easily paid for, is there another item of similar value that could be given (to the victim)?*
- *Is there an item of similar sentimental value—that is, something the child really cares about and values—an item that means a lot to the child?*
- *Would it be reasonable for the victim to explore other possessions that the child has which might be a fair exchange or trade to replace the item that was lost in a fire?*
- *If no object can be paid for or replaced, would it seem reasonable to ask the child to do a "work task" for the person whose property was damaged (*e.g., *cleaning up the house or yard, organizing materials in a few rooms, putting things away, running an errand)?*

Help the caregiver develop a restitution plan that identifies the item to be exchanged/replaced or the work task to be completed, and then review some rules for conducting it:

- It is important to explain to all parties why this is happening and how/when it will happen.
- The item exchange/replacement or work task should be as fair as possible.
- The child should understand and agree to what's going to happen:
  - "Because you burned Joe's book, which he can no longer read, you will need to replace his book."
  - "Because your fire made the family room very smoky, you will need to clean the rug, curtains, and the furniture to eliminate the smell of smoke."
- The exchange/replacement or work task should be done in a matter of fact way, and not with any vindictive or angry tone.
- The child should be praised for participation.
- Remember that this could be an emotional or challenging event for the child who set the fire, and possibly the victim of the fire as well.

The parent may also want to set up a similar restitution plan (contingency) for a future incident when fire is misused and leads to some type of damage or debris, however minor these outcomes might be (e.g., smoke damage). Here, the parent would simply identify what is the restitution consequence the child would experience following a new incident. This plan may help to motivate the child to avoid using fire and causing any damages to property of any kind.

Identify Potential Target Behaviors to Decrease

Show the "Possible Problem Behaviors to Discourage and Consequences" handout (Parent Form 11.3) found in appendix C.

Earlier, we identified some positive behaviors to be targeted. And we've also discussed some problem behaviors related to the misuse of fire that may be important to target, so let's review the list and then decide which of these problem behaviors you would like to target using effective consequences. Sometimes, the combination of both strategies is more effective than either one alone. This form is set up in the same manner as the "Positive Behaviors to Encourage" handout (Parent Form 8.9), except the "Problem

Behaviors to Discourage" handout identifies the child's negative behaviors, and has lines for adding the corrective or negative consequences that might be used to discourage these behaviors.

Here are some examples of a child's problem behaviors one could target:

- *Keeping/hiding/having matches*
- *Matchplay, fireplay, burned materials/burn marks*
- *Community report of the behaviors listed*
- *Inability to describe activities with peers in detail to parents*
- *Wandering/absence or discrepancy during a monitoring check*

Ask the parent to identify a few specific problem behaviors to target and then decide which ones to include on the form.

Discussion of Home Management (Consequence) Program

Rationale for Combining Positive and Problem Behaviors in One Program

(Child's name) and I have talked about steps for solving problems in ways that are positive and helpful, rather than negative and dangerous. We encourage parents at this stage to think about ways that they support their children/adolescents to solve problems effectively. One way to do this is by using rewards or attention. Most of the children/adolescents we work with like to earn privileges or receive things as they start to behave appropriately—it helps them understand that their improvements are good and to be encouraged. We talked before about some of the methods that you can use to motivate (child) to show this kind of special behavior. And, in our discussion today, we discussed some possible problem behaviors we may want to target using different consequences. There are different strategies that parents can use to help children/adolescents improve their behavior. Let's talk more about the different ways you can do this that would be best for your child.

Identify the desired behaviors and rewards the parent wants to consider for a program at home. Help the parent generate more ideas if necessary. Try to make sure that each desired behavior represents a specific domain (e.g., cooperating, acting responsibly, showing good emotional control) to avoid having lots of behaviors that are very similar to (and possibly redundant with) one another.

- Potential Positive Behaviors: *"What are some of the things that you want (child) to do, and some of the rewards you might be able to use?*
- Potential Rewards: *"What rewards could be added as a special program for home?"*

Identify one or more problem behaviors and some consequences (e.g., lost privilege) the parent wants to consider for a program at home. Help the parent generate more ideas if necessary.

- Potential Problem behavior: *"What do you think should be included in a home program that you're really concerned about?"*
- Potential Corrective Consequences: *"What corrective consequences could be added as well?"*

Explain the Basic Elements of a Home Management Program

*Let me tell you about some home programs that you can use to change those behaviors we just discussed. One is called a point system. A point system allows children/adolescents to earn points—or some other symbol (*e.g., *token, chip)—after completing specific behaviors. These points can then be used to "purchase" privileges, tangible items, or special activities. A system like this can be useful when parents are looking to target 3 to 5 behaviors. You can also include target behaviors that you want to discourage or decrease, which can lead to a loss of points earned. This type of program usually appeals to children and adolescents of all ages.*

Now, you can use a similar program to manage the child's behavior that enables her to earn an allowance or money, usually for completing chores and other tasks. An allowance program allows the child/adolescent to earn money for completing chores or tasks. This type of system can be helpful when parents are looking for the desired child behavior to occur once or twice a day. Because the reward is money, it often appeals to older children and adolescents who value money and have opportunities to spend their allowance. You could also exchange points for a certain amount of money or just specify how much money is earned for each positive target behavior. Of course, an allowance program can also include negative target behaviors that could lose the child money each time he does them. You can think about which one might be better to try first.

Point Systems

Review Parent Form 11.4 (Point System Program) located in appendix C.

If you use a point system, (child) could earn (or lose) points for completing certain tasks, following certain rules, or for simply avoiding doing things you don't like. This method may help her become more involved in behaviors you support or encourage and less involved with (interested in) behaviors you want to discourage, like playing with fire.

Review all points, clarifying each one as needed.

Let's review this form to learn what has to be done in order to have a good point system.

Guide the parent on how to use the steps noted in the handout to develop a point system.

Practitioner Note

If using points is too complicated or simply not necessary for a particular child, the parent can just use hatch marks (////) to record the frequency of each behavior. The parent would then identify the number of incidents of a given behavior that would result in a specific consequence.

Potential Challenges to Implementing a Home Management Program

Review Parent Form 11.5 (Potential Challenges to Implementing a Home Management Program) with the parent. This form is found in appendix C.

"Forgetting" to follow the plan:

Regardless of your mood or what has just happened that minute or that day, you have to be consistent in giving or removing points or privileges. Whatever your child/adolescent has coming should be done.

Setting up a system that is too extensive:

We don't want to set something up that is too big or hard to manage.

Offering a reward or privilege that is too big:

Really big rewards aren't that helpful because they can't be used that often. Instead, try to use small to moderate ones. What would you use as a small reward for you child? What would you use for a large reward? How feasible are each of those rewards?

Not having a desirable reward or privilege:

Sometimes it is hard to find an effective reward, one that (child) will really work to get.

Poor monitoring of performance:

Monitor outcomes and be consistent, regardless of the outcome.

Home Practice

Ask the parent to review Parent Forms 11.4 (Point System Program) and 11.5 (Potential Challenges to Implementing a Home Management Program). Suggest that the parent begin to develop a basic contingency program that outlines specific behaviors to be targeted and the consequences that will be administered following those behaviors.

Please take a more careful look at this Point System Program handout (Parent Form 11.4) and the Potential Challenges to Implementing a Home Management Program handout (Parent Form 11.5) to decide which behaviors you would like to encourage or discourage, and what consequences—rewards or correction—might be effective.

Being Assertive: Standing Up for Yourself

MATERIALS

- "Being Assertive" (Child Form 12.1)
- "Requests" (Child Form 12.2)
- "Ways People Stand Up For Themselves" (Child Form 12.3)
- "I and You Statements" (Child Form 12.4)
- "Complaints" (Child Form 12.5)
- "Making Complaints" (Child Form 12.6)
- Pencils, pens, and blank paper

OBJECTIVES

- Teach the child about assertive behavior, in contrast to aggressive or passive behavior.

Introduction to Being Assertive

Explain the rationale for why children should be assertive. You may say something like:

Today we are going to talk about some things you can say to help you ask for things you want or need, and to stand up for yourself. Some kids/teens are good at saying or doing things that tell others what they would like them to do, or what they want someone to stop doing or not do. Those kinds of statements can help make things happen that you care about or want.

Talk about the importance of making requests the right way:

It is important to know how to ask for things in an assertive way. Everyone asks for favors. For example, you might ask someone if you can borrow a pencil, or to help you, or to give you information. Now, there are different ways to make requests. If you make a request in a nice way, then you might have a better chance of getting what you want. For example, you should ask for something nicely rather than ordering someone else to do something you want.

For example, you ask your friend to help you with your homework:

- *Wrong way: "You better help me with my homework."*
- *Right way: "Will you help me with some of my homework? I don't understand some of it."*

The difference is that, in the second example, you asked in a nice way. Also, your friend knows what you want and how important the favor is to you.

Use these probe questions for further discussion:

- *What kinds of requests do you usually make?*
- *Give at least two different ways of making one of these requests.*
- *Which is the better way to ask? Why?*
- *How do you feel when someone asks you in a way that is pushy or loud?*

In-Session Practice

Have the child write down one situation where he could have been more assertive in the past week. Use Child Form 12.1 (Being Assertive), located in appendix B at the end of this Guide.

I'd like you to think of at least one situation that happened at school, at home, or anywhere else where you wish you would have been more assertive, like where you think you should have made a request or asked for something from someone else.

Ask the child:

- *Describe the situation.*
- *What did you do and how did you feel?*

- *What got in the way of acting assertive?*
- *What could you have said?*
- *What would have been the best way to solve the problem or to be assertive?*
- *Was there anything that you did that you wish you hadn't done?*

Show Child Form 12.2 (Requests), found in appendix B:

- *Use a neutral tone of voice (not loud or hard to hear).*
- *Ask, don't demand.*
- *Be simple and specific about what you want.*
- *Give a reason for why you are making this request.*
- *Use "please" and "thank you" to be polite.*

By making a request in a nice way, it is more likely that someone will honor it. If you're angry, you make someone less likely to want to honor that one and any other ones you make later on.

Model some example of ways that the child could ask for a favor in the right way:

- *You have just realized that you forgot to bring your lunch to school. You are going to have to borrow money from a friend.*
- *You didn't take a basketball to the gym and want to use a friend's.*
- *You want to use someone's phone.*

Practice Situations

Using Child Form 12.2 (Requests), practice generating effective requests. As needed, consider using role-reversal to model how this could be initiated (where you serve in the role as the child to model the behavior):

- *You are with some friends and would like to buy a snack. You just noticed that you didn't bring any money.*
- *You want to ask a classmate for help with a hard homework assignment.*
- *You just found out that a friend has a toy/book that you really want to play with/look at.*
- *You would like to invite a friend over for dinner. You are asking your mother for permission.*
- *You need help fixing your bicycle and want to ask your dad for help.*

- *You don't understand the homework assignment the teacher gave you. You walk up to her after class and say…*
- *Think of some things or activities that you would like to do at home or with friends and practice asking to do these things.*

Practice Standing Up For Yourself

Now, let's talk about standing up for yourself. That means that you let other people know what you want them to do, like helping you to do something or giving something to you. And, it also means that you let them know what you do not like or want them to do, like asking someone not to call you a certain name, but saying this in a way that makes it easy for others to listen to and respect what you said. So, being assertive like this means that you are able to stand up for yourself, but without using words or actions that might make others feel badly or hurt. Saying the right thing and keeping your cool while saying it can sometimes be a real challenge. So, that's why we'll talk about how to do this well.

Now, some of the time, people won't know that they are doing something that you don't like. For example, someone might borrow your pencil in class without asking, or you have a bunch of friends over and they leave without helping clean up a big mess. If you don't stand up for your rights by telling them what you want, people might take advantage of you and it might even happen to you again because someone thinks it's okay.

Consider using some probes if further discussion for being assertive is needed:

- *Is it important for you to stand up for your rights? Why?*
- *What would (could) happen if you don't?*
- *Can you give me an example of a situation where it would be a good idea to (or, you should) stand up for your rights?*
    - *What would be your rights in this case?*
    - *What would happen if you don't say or do anything?*
    - *What would be the best thing to say in this situation? Why?*

Three Ways People Stand Up For Themselves—Assertive, Passive, and Aggressive

There are helpful (good, appropriate) and not so helpful (bad, inappropriate) ways for children or adults to stand up for themselves. We do not want to encourage being passive, like not doing anything to change the situation (e.g., just ignoring or avoiding the person), and we don't want to encourage being aggressive (e.g., getting out of control, using threats, or using force).

Explain good and bad behaviors using Child Form 12.3 (Ways People Stand Up for Themselves) located in appendix B. For more background, see these sources: Deluty (1979) and Michelson et al. (1982).

Let's talk about some different ways in which people try to stand up for themselves. Some of these ways aren't so good or helpful—that's when people are either passive or aggressive. But, when people are assertive, they seem to get better results. Let's talk about each one:

1. Being *Passive* (the "Mouse")

Definition for Practitioner

A person who is passive usually avoids asking for things or making his needs known to others. That is, this person may not be assertive when he needs or wants something. Since the person doesn't ask or make a request, his needs, feelings, opinions, or rights may be ignored or simply not recognized. Others may take advantage of him or control him. Here the person allows his feelings to be disregarded. If he does ask, his tone is often quiet, shy, and afraid. This kind of behavior is passive—like being a mouse, which is usually small and scared, so it often hides from or avoids others.

Definition for Child

This is when people don't express themselves. They ignore their own opinions or feelings, so others often take advantage of them. They may also let others boss them around or say things that are upsetting or mean, and don't stand up for their rights.

2. Being *Aggressive* (the "Monster")

Definition for Practitioner

An aggressive person is one whose behavior ignores or denies the feelings, opinions, needs, or rights of someone else. The person takes advantage of others and tries to control them, or prevents others from expressing their feelings or doing something. He may even threaten to hurt someone. The tone is bossy, demanding, pushy, and intimidating. This kind of behavior is seen by others as mean—kind of like being a monster. A monster doesn't care about others or how others feel about them, so he may be willing to do anything he wants to get something, regardless of how that affects someone else.

Definition for Child

This is when people pick on others and put them down; they only care about getting what they want, when they want it. They don't care about how others feel and often get into trouble because of this. They take advantage of others. They might stand up for their rights, but they hurt other people in the process. Sometimes, these are people who use physical force to control another person.

3. Being *Assertive* (the "Me")

Definition for Practitioner

When you're assertive, your behavior recognizes the needs, feelings, and rights of yourself and another person. Here, someone tries to do something positive that might be an acceptable way to find a solution for that person and someone else. Doing something like that may also help to keep their relationship positive. Here, the person expresses himself/herself without putting others down or allowing his/her own rights or feelings to be violated or denied. The tone is honest, positive, fair, considerate, sensitive, and helpful.

Definition for Child

This is when people let others know how they feel in a way that doesn't hurt or bother others. You're not hurting other people, but you are still helping yourself and standing up for your rights. Another example is when you respect others' feelings and you don't force yourself on them. You let others know your feelings without ignoring their feelings.

Describe Some Situations for Illustration

Give the child some examples of responses and have the child guess whether each response is passive, aggressive, or assertive.

Let's talk about some of the situations where it would be good for you to stand up for your rights. I am going to give you the situation and three possible responses. Then, I would like you to figure out which response is passive, aggressive, or assertive.

Situation 1

You and your friend are playing a board game. Your friend says to you: "You can't do that! I don't think you can move that way."

- *Passive: "But I thought... well, never mind." Then you put your piece back.*
- *Aggressive: "Either you let me move this way, or I'll move your piece all the way back to start."*
- *Assertive: "I think I can move that way, but let's check the rulebook to see."*

Situation 2

You are in a store looking for a gift for a friend. One of the store clerks comes up to you and says: "If you're not buying something, you are going to have to leave."

- *Assertive: "I am buying a gift for my friend."*
- *Aggressive: "What's your problem?"*
- *Passive: "Ok, I guess I'll leave." Then you leave without buying a gift.*

Practice the Situations

Present the following situations, and ask the child to offer examples of the different types of responses (assertive, passive, or aggressive).

Let's discuss some more situations. I'd like you to tell me what you would do in each of these situations; give me an example of a way to handle the situation that is assertive, passive, and aggressive; and tell me if this situation or something like it has ever happened to you.

- *Someone borrowed an iPod from you a while ago. You've already asked for it back once.*
- *You just bought an ice cream cone and you got the wrong amount of change.*
- *You are standing in the lunch line and someone cuts in front of you.*

- *Your sister tells you that it is your turn to wash the dishes, but you washed them yesterday so it is her turn.*
- *You've been watching a movie on TV when your mother/father walks in and changes the channel without saying anything.*

In-Session Practice

Have the child write down one situation where he could have been more assertive in the past week. Use the same Child Form 12.1 (Being Assertive) from the previous section.

I'd like you to think of at least one situation that happened at school or at home where you wish you would have been more assertive, like where you think you should have stood up for your rights differently.

Ask the child:

- *Describe the situation.*
- *What did you do and how did you feel?*
- *What got in the way of acting assertive?*
- *What could you have said?*
- *What would have been the best way to solve the problem or to be assertive?*
- *Was there anything that you did that you wish you hadn't done?*

Practice How to Respond to Everyday Hassles and Complaints

Introduce this next section by transitioning away from the previous section:

Now, what should you say if you don't like something that someone did or said? The best way to handle this situation is to let the person responsible know you are unhappy. You should express yourself in an assertive way without hurting that person or making her feel bad. Making a complaint in the right way (assertively) doesn't guarantee that you'll get what you want, but it usually increases the chances that you will. Why do you think it is better to let others know what is bothering you in an assertive way, instead of hurting them or making them feel badly? Are there some disadvantages to making others feel bad or hurting them in some way? Does it seem useful in the long run to use force to make someone do something?

- Get the person's attention in a nice way (use name, say "excuse me").
- Explain the situation and describe what happened.
- Tell the person what you didn't like, using "I" statements. Refer to Child Form 12.4 ("I" and "You" Statements), found in appendix B.
  - Example: From "Why did you do that?! You're so mean!" to "I don't like it when you do that, it makes me feel bad."
- Don't yell or get out of control.
- Remember your tone of voice. Keep it calm and cool, and avoid sounding angry or sarcastic.
- Listen to what the other person has to say.

Using Child Form 12.5 (Complaints), try role-playing with the child so he understands how to let another person know when something is bothering him:

Let's try some role-plays where you might want to act assertively. Then, we can talk about how you handled each situation.

Practitioner Note

It is helpful to reverse roles for some of the first situations, so that the child can see how you would handle them properly. If the child is having difficulty understanding the difference between passive, aggressive, and assertive responses, also role-play some aggressive and passive complaints. Practice as many/few as are necessary to convey/model the concept.

- *Your friend returns a book she borrowed and the pages are wrinkled from something she spilled on them.*
- *Someone has said something that hurts your feelings.*
- *Kids sitting behind you are making it hard to hear the teacher.*
- *You see a friend making fun of one of your classmates.*
- *You are late for a movie because one of your friends arrived late.*
- *You overhear a friend saying some bad things about another child.*
- *You learn from one friend that another friend didn't invite you to a party.*
- *You're playing basketball and your partner won't pass.*

- *Your friend returns your iPod that he borrowed and the screen is scratched.*
- *You're working on a school project with a friend who isn't doing any work.*

Points to Review About Complaints

Explain that a positive, assertive complaint can make a situation better. However, an aggressive, angry complaint usually makes a situation worse. An aggressive complaint does not make the other person want to change. Usually, both parties then feel frustrated. Explain that the child should complain only when it is necessary. Tell the child that she should not use an aggressive complaint just because the other person does not seem to be listening to her positive, assertive complaint.

In-Session Review

Using Child Form 12.6 (Making Complaints), review with the child how to effectively make complaints:

Try to think of something that bothered you the past week and that you should have complained about using an assertive complaint. Did you see someone else, like one your friends or one of your parents, make a complaint? If yes, what did you notice about how they made their complaint? How did the other person respond to them?

Home Practice

Assign a brief home practice task that will help the child practice a skill in the real world. Decide how challenging to make the situation (simple or complicated?) and whether it would be useful to include a particular person. It could easily involve using the skill in two different settings, or something more specific based on the child's feedback during the session.

Update and Review of Home Management Plan/Program

MATERIALS

- "Home Management Program (Contingency Plan)" (Parent Form 13.1)
- Pencils, pens, and blank paper

OBJECTIVES

- Develop and/or finalize the family's home contingency program.
- Discuss the final details of the program.

Home Behavior Management: A Brief Review

Rationale

Most of the material on the use of consequences (i.e., contingencies) to manage and support the child's behavior at home has already been discussed. So, this chapter includes a review and extension of this information in order to help the parent finalize and implement a home management plan that is both feasible and effective. Specifically, you will want to help the parent select the target behaviors and consequences that will be used from among the various options discussed in prior sessions. Explain to the parent why it is important for him to create a plan to make clear the rules and expectations of the household. It is helpful to explain that all families that go through firesetting behavior interventions do this to set the boundaries within the household.

Discuss Any Current Plans or Programs that Target the Child's Behavior

Ask the parent to describe and then review any current behavior management plans (contingencies), how well they have been working, and any concerns or problems with the plan.

In previous sessions, we have talked about using some different strategies to encourage (child's) good behaviors, and to discourage the ones that may cause problems. Let's talk about each behavior you are trying to target (positive and problematic ones) and the strategies you're using now. Let's first talk about any behaviors involving fire. Then, we'll learn the consequence you administer after each behavior. Finally, we'll discuss any concerns or the need for updating the plan.

Here are some examples of fire-related child behaviors the parent may report:

- Keeping/hiding/having matches
- Matchplay, fireplay, burned materials/burn marks
- Community report of the behaviors listed above
- Inability to describe activities with peers in detail to parents
- Wandering/absence or discrepancy during a monitoring check

Here are some potential reward-oriented consequences for these behaviors:

- Access to events, activities, and privileges (e.g. earning game time, use of telephone or electronics, social activities with friends)
- Access to tangibles (e.g. earning a gift card for music or a download)

Here are examples of positive behaviors that the parent may report:

- Giving the parent any incendiary (or other unsafe item in the child's possession)
- Reporting to the parent any activities with peers
- Telling the parent where the child will be (when supervision is needed)
- Being more involved in appropriate individual/peer activities

Here are some examples of consequences that may target these behaviors:

- Earn an extra dessert or 30 minutes of TV time
- Earn access to a special event or peer activity
- Earn a later curfew
- Earn more unsupervised time

Here are examples of program challenges or concerns:

- The parent has a hard time monitoring the behavior or following through with consequences.
- There is no change in the child's behavior.
- The program is too difficult to maintain.
- Other targets or consequences need to be added to the program.

Revise the Program with Parental Input

Ask the parent to make any revisions to the program by identifying any new targets or consequences for the program. If helpful, you can review Parent Form 5.3 again for a list of different problem behaviors (Behaviors Associated with Children Who Play With or Set Fires and Some Possible Consequences for Managing These Behaviors) located in appendix C. Ask the parent to list all behaviors that need to be targeted and discuss the specific consequences considered for each behavior.

You can also review Table 13.1, which provides an example of a plan with some reasonable targets and consequences designed to decrease various

Table 13.1 Sample Problem Behaviors and Reasonable Consequences

Problem Behaviors	Reasonable Consequences
Incendiaries found on child/adolescent (or a community report of this)	Privilege loss
Fire/match play (or a community report of such activities)	Restitution—clean up task and item exchange Privilege loss If with peer, brief loss of privilege to be with peer
Absence/wandering or discrepancy during monitoring check	All time now supervised Earlier curfew

problem behaviors. By this time, it is expected that the parent has already identified at least one positive behavior to increase, but, if not, please identify and then define a positive behavior to target.

Review the Updated Home Management Program

Once all of the behaviors and consequences are identified, use Parent Form 13.1 [Home Management Program (Contingency Plan)], located in appendix C to help the parent complete the "final" plan and make sure she can explain it to the child. If the family has already implemented a behavior contract or allowance system, use whatever parts of that system are still appropriate and needed.

Here is a home management program form for you to use with your child. It will help assure that both of you have the same understanding of the new rules at home. It is useful to identify the details of the program you plan to administer, and to review these details with the child in advance to ensure that the program is clear, understood, and fair. If changes are suggested or made, simply change the language as needed.

- *How the does the plan sound?*
- *Do you feel comfortable explaining it to (child)?*
- *Do you want to add a special bonus reward if (child) shows exceptional gains in several of these positive behaviors?*
- *Are there any special stipulations or conditions that need to be met or followed in this program?*
- *Do you know what specific responsibilities or activities you will complete to make sure the plan is successful?*

Review the Home Program the Family Wants to Apply

Before finalizing the home contingency program it is important to make sure that the parent is comfortable with the plan and understand all aspects of the plan. Soliciting feedback regarding the home contingency program allows for feedback about how the plan ran in action since the last session. If any changes are needed, the program can be modified accordingly, and any ambiguous details can be clarified. After the program update is finished, any ambiguity or problems with the plan may need to be modified by the parent. Take the opportunity to model the appropriate ways to handle problems with the plan, and resolve them so that the parent and child can do so together.

Make sure that these details are addressed to structure the specific home contingency program:

- Determine and write down any problem behaviors that will be targeted along with the consequences that will be paired with them.
- Determine and write down the contingencies the family will use.
- Discuss and write down appropriate parent responsibilities and any consequences that may occur if these responsibilities are not met.
- Discuss and write down any special stipulations that seem appropriate.
- Answer any questions that the parent or child/adolescent may have. Procure parent and child/adolescent signatures and make two copies of the contract (one for the parent and one for the child/adolescent).
- The family should then post the contract in a highly visible location in the home so the child/adolescent can view her progress regularly.
- The parent should review the home program regularly with the child—follow-through is important.

Emphasize the importance of consistency when delivering rewards and punishments.

Do your best to follow the home program as it is written. Try to discuss any disagreements or problems you have and write them down so they can be reviewed later on.

Ask the parent to anticipate any problems with the agreement or program. If he does, help him proactively create some solutions ahead of time. Also ask the parent what obstacles the child would identify with the plan.

The three major tips that you want the parent to leave with are:

1. Post the home management plan so that it is visible for frequent checks of progress, and to review the progress frequently.
2. Remember the importance of being consistent in delivering earned consequences for positive and problem behaviors, so the child knows that the parent is committed to the program.
3. Follow the plan exactly as it is written, and discuss any disagreements with your child, or during a follow-up conversation/session (if necessary).

Termination Preparation

After the contingency plan is made, the next session is the termination session. Ask the parent to return with the child, and to bring all materials they wish to review. Informing the parent (and the parent can inform the child) that termination of the firesetting behavior sessions is near gives them the opportunity to identify any other concerns or to ask clarifying questions about any of the materials from any of the sessions. If necessary, additional sessions can be scheduled before the termination session.

Home Practice

Review, explain, and then use the home management program developed in this session. Report in the next session any feedback or suggestions you want to make to the program.

Skill Review, Follow-up Plans, and Termination

MATERIALS

- "Program Completion Certificate" (Practitioner Form 14.1)
- Pencils, pens, and blank paper

OBJECTIVES

- Explore skills that have been taught and used.
- Discuss potential problem situations and plan for them.
- Determine need for referral.
- Complete graduation routine.

Review Any Home Practice Assignment From the Prior Session

Ask the parent if she has developed or implemented a home management program (contingency plan) and, if so, ask about the targets, outcomes, and any feedback on its use. Consider whether the plan needs any revision to be successful.

If the child is present, ask the child about the same program, and solicit similar details to learn about his experiences with the program. As needed, modify the structure or details of the plan based on their feedback to enhance the likelihood that it achieves the desired outcome and that they will continue to implement it.

Review the Use of Individual Skills and Family Programs During Treatment

With each participant, summarize the primary skills that have been discussed throughout the treatment.

Since today is our final session, I'd like to take a few minutes to review some of the things that we've talked about in treatment. You've learned so much over the past months/weeks! Which skills do you remember talking about here? [Once both are done, continue here]. *Ok, good. So, here are the main skills we've learned. Let's briefly review the key points in each one.*

- *Understanding the child's recent fire incident(s) and any specific behaviors to target*
- *Home fire safety skills*
- *Basic parenting skills (attending, ignoring, praise, and rewards)*
- *Recognizing and managing our emotions and thoughts (relaxation, automatic thoughts)*
- *Effective problem-solving (five skills)*
- *Parent management (natural and logical consequences, removing privileges, time out)*
- *Being assertive and standing up for yourself*
- *Using a home program (targets and consequences)*

Identify Specific Applications of These Skills That the Child and Parent Have Used

Begin with the parent and then the child. Make sure that important skills or procedures that were used successfully are discussed. Try to identify at least two applications for the parent and child to discuss.

Please describe what skill you have used, and how and when you might use it again.

Discuss the Benefits That Have Resulted From Their Use

Explore different types of benefits or consequences they have noticed—for the parent, child, family, and/or any others involved in the problem.

Discuss Any Challenges to Using the Skills and Any Potential Changes Needed

Please let me know if we need to change or tweak any tool used on your plan. [Allow the parent and child to review and discuss a revised plan for using a given skill.]

If not already covered above, explore some specific skills or behaviors that were designed to avoid any use of fire or controlling access to fire materials. Reinforce these skills, and offer any other suggestions to enhance or ensure their usefulness.

To child: *What tools have you used to avoid being around things or people that cause fire?*

To parent: *What tools have you used to make sure that no fire materials are available to start a fire?*

Ask them about the following safety issues:

1. *Are matches, lighters, and other devices kept in a secure and safe place now, away from the child?*
2. *Does the child know where there are, and can she get to them?*
3. *Are they both committed to keeping their home "fire-safe"?*

Ask Both the Parent and Child About Their Plans to Continue Using These Skills

Support both the child and parent in their efforts to use the skills and their commitment to using them in the future. Remind them that they now have the skills to make thing better—they just have to choose to continue to use them. It's just like having a seat belt in the car or a warm coat—if you use them, you stay safe and warm.

Solicit any final thoughts about the program from the parent and child, if they are willing to share. Then, praise and congratulate the family on completing the program and making the progress that they did.

Explore the Need for a Safety Plan or Outside Referral

Identify and Address Any Potential Child/Family Problems (e.g., Family Safety, Security, Stability)

Remind the family that it is important to plan for how they will address future problems and to problem-solve these issues together. Ask them to identify any current or potential problem situations that the child or family may encounter in the next few weeks that could increase the child's risk of misusing fire. Ask them to report and discuss even a minor upcoming concern—can they imagine any problems that might arise in the near future?

As we prepare to finish our work together, are there any other problems or issues that you'd like to discuss?

Brainstorm Any Potential Future Family Problems

If something is mentioned, discuss with the family relevant skills from treatment and problem-solve together how to solve potential issues.

Okay, help me understand what you mean here (what they are) and we'll use some of your tools to make sure they are addressed.

If nothing is mentioned, consider raising a few possible situations based on your knowledge of the family's current status. List a few situations or issues as relevant, and ask how they would be addressed. Some examples may include:

- Child being left home alone
- Child being with/spending time with friends who use lighters
- Child and sibling cooking on stove
- Caregivers leaving lighter in basement

Apply Problem-Solving Steps to Each Situation in Order to Develop a Reasonable Solution and Then a Plan to Use It

If a safety plan is needed to further reduce risk or harm, carefully identify what the parent needs to do based on what the child does.

Let's use the problem-solving steps to identify a useful solution or response here. What's the problem?..... now, what's the goal?.... ok, and what about some possible solutions—what could you do here?.... what are the pros and cons of each option?... And, which one do you want to try first?

Confirm any solution and the plan to put it in action.

Prepare for Termination

Encourage Persistence in Using New Skills

If you say, "I think you are so much better," to a family who may be resistant to termination, they can respond by trying to prove to you that they are not really better. So avoid global comments and instead provide reinforcement about specific behaviors/events.

Explore How Difficult It Might Be to Use New Skills Instead of Old Ones

You have been using lots of new tools, and there will be times when you are going to want to go back to doing things the way it used to be because it's easier. Let's talk about how and what you can do to continue to remind yourselves to use the new tools as often as possible.

Discuss Any Requested or Recommended Referral Options

If you are uncertain about referrals for additional services, ask the parent if he sees the need for continued services. Explore whether a referral is needed to address the child's referral concerns or other issues related to the child's behavioral health needs.

If a referral seems needed, discuss what information will be provided to the referral source regarding the family's work in the treatment and the likely reason for the referral.

Practitioner Note

Depending on the specific child/adolescent, you may decide to have this conversation with only the parent or with both the parent and child/adolescent.

I will forward a treatment participation summary with some basic information in it, such as participation, attendance, and general content. Is there anything else that you'd like me to include?

Discuss with the child/adolescent and parent what to do about any crises or emergencies.

If there was an emergency or crisis, who will you call and for what?

Where will you go?

What will you do?

Check-in about any "next steps" that are unclear or if there is anything else the family needs.

Do you have any questions about what is happening next? If questions come up you can contact me, and I can do my best to answer your questions and/or refer you to someone who can help. How does that sound?

Encourage the Family to Access any Needed Community Resources

As noted, remind the parent and child/adolescent about calling you if they need special assistance, have questions, or need to return to therapy. If they have an immediate fire-related crisis, recommend that they call the local fire department or 911. Offer any other information about other community resources that might be useful to the family, including other service providers, local community programs, and online resources for parents.

Graduation Routine/Recognition

Graduation Rituals

Commemorate the graduation. To mark this occasion, consider giving the child and family a certificate or other memento acknowledging their participation in the program, or some other symbol that they can keep. It matters less about what you give the family—and more that you acknowledge their accomplishments and gains.

Giving a certificate to the child is a way of showing that he learned important ways to keep himself and the community safe. Parents may also benefit from participating in this part of the session to both support the child and indicate how they themselves have also benefitted from the program. A sample certificate is found in Practitioner Form 14.1 in appendix B at the end of this Guide.

Offer Some Final Comments

Offer a brief summary of the family's achievements and status, highlighting where there's hope for recovery and maintenance of positive gains/relationships.

Provide a positive "send-off" message, including Practitioner Form 14.1 (Program Completion Certificate).

Professional and Program Development

MATERIALS

- "Sample Consultation Report with Recommendations" (Practitioner Form 15.1)

OBJECTIVES

- Explore possible options for promoting professional development.
- Consider ways to enhance program capacity and sustainability.

Professional Development

Education and Training

The field of juvenile or childhood firesetting has grown steadily since the U.S. Fire Administration first developed and disseminated materials to support an organized community-based program for this population in the late 1970s. Any professional can now find numerous materials in print and online dealing with numerous topics of relevance to working with children who set fires and their families. Some of these topics are highly specialized and convey the advances made in the various technologies available to combat the misuse of fire. These advances are needed to ensure that adequate methods are available to prevent and suppress the dangerous use of fire. Many of these materials are available through numerous organizations and service sectors, and they are often written by and for diverse disciplines (e.g., fire service, medical and burn care, mental health, social services, and law enforcement). In addition, there are far more professional trainings in this field now than even 10 years ago, many of which are sponsored by FEMA/USFA, the offices of the state fire marshal or fire commissioner, and individual professionals. Keeping up with advances in the field is a critical step toward maximizing the likelihood that emerging methods and technologies are being used correctly and effectively. We suggest that these diverse offerings be considered on a routine basis to enhance skill development and build intervention capacity. We encourage all programs to commit to the preparation, training, and ongoing monitoring of educators or practitioners who routinely apply these interventions. Such efforts would promote more large-scale dissemination of new material and encourage the use of more effective strategies based on program evaluation or more rigorous outcome data.

Consultation

In addition to having access to new educational and training materials, we also recommend that practitioners working primarily on an independent basis consider being involved in some form of peer or professional consultation. Given the complexity of these cases and the broad array of methods/materials that this book and other resources (e.g., Kolko, 2002) emphasize, it seems prudent to consider having an opportunity to review one's cases, discuss specific implementation challenges, and obtain outside advice and consultation within a small peer group and/or with a local or regional expert consultant. As noted in prior sections, there are many challenges to working with children and adolescents who have misused fire that might be more easily addressed if consultation feedback is available. Further, novice practitioners often request this type of supportive feedback and advice, which may help to minimize burnout and the stress of serving a population whose behaviors have the potential for personally dangerous and/or professionally threatening consequences. In our work, we routinely conduct peer supervision with our teams and maintain contacts with other professionals to ensure that all of us are learning from one another, applying sound and well-justified intervention strategies, and taking all reasonable precautions to ensure community safety.

There are also times when receiving or providing consultation to local programs can provide an important resource to the community. Practitioners

with expertise in this field are an important resource for conducting specialized programming related to firesetting assessment, prevention, intervention, consultation, and program evaluation or research. Effective consultation could include general recommendations and a written report, through program agreements that specify the tasks and a modest fee. Practitioners can be engaged to conduct and summarize the assessment in coordination with a local program and the family. Some of these specialized functions could include the following:

1. Review of all clinical records from service providers (as permitted) to include all treatment information including, but not limited to, a psychiatric evaluation, treatment plan and notes, discharge plan, and any additional notes relevant to the identified child.
2. Review of fire reports from the fire marshal, arson report, cause and origin report, and any other salient information regarding the specific incident.
3. Discussion of logistics and other details/issues to ensure relevance of assessment and treatment and family readiness for fire-specific treatment intervention.
4. Conducting a formal assessment using specific fire assessment measures and interviews with both the child/adolescent and the parent/caregiver.
5. Preparation of documents for entry into a medical record or for submission to a legal or law enforcement officer (e.g., family court judge, probation officer) or other authority (e.g., supplemental treatment notes after each assessment session).

Program Development

Screening and Assessment

Given the importance of assessing several activities related to a child's involvement with fire, it is important to maintain an understanding of any advances in the procedures recommended for assessing children who misuse fire and their families. Although there are very few such instruments with sufficient psychometric properties, measurement developments may be reported in publications or websites affiliated with the fire service and mental health professions. One may find updates in the evaluation of several domains related to the misuse of fire and chemicals having pyrotechnic effects (e.g., firesetting, matchplay, bomb-making) and other fire-related activities, including how to capture one's interest in or attraction to fire.

Intervention Services

As a reminder, the use of FSE and CBT showed parallel and greater improvements in reducing involvement with fire than a home visit from a fire fighter (Kolko, 2001), including the frequency of firesetting and proportion of children playing with matches, severity of individualized problems with fire, and involvement in fire-related acts and other deviant fire activities. These and other group differences remained at 1-year follow-up. Therefore, much of the content in this guidebook reflects the use of interventions that reflect contrasting conceptual models—that is, fire safety education and cognitive-behavioral treatment. In general, the selection of our materials was based, in part, on both practical and empirical considerations that support the use of this content. Many intervention programs or curricula emphasize one or the other, based on the primary technical expertise of the staff in the program. Our combination of complimentary material hopefully integrates the key content and ingredients from both topical areas to enhance service impact on a range of positive and problem behaviors. Of course, new developments in intervention content and methods are likely to be reported and hopefully subjected to program evaluation to determine their utility. We suggest that multipurpose programs continue to integrate educational and clinical approaches to intervention with children who have used fire inappropriately.

Evaluation and Research

Programs or providers working in this field are encouraged to contribute program evaluation or research study data to enhance the use and impact of our assessments and interventions. Novel measures that reflect more current mechanisms for causing fires (e.g., bombs/chemicals) or that more specifically relate to fire recidivism are needed to guide the next generation of practitioners. Further, we need replications of efficacy studies to determine whether comparable outcomes are produced by these interventions when conducted by other providers and with different populations.

It is also important to develop and test adaptations of these approaches for other populations, such as younger children or adolescents, children with developmental or intellectual limitations, and children in residential or other unique settings. The materials or content upon which these interventions are based requires further refinement and articulation. For example, there is a broad array of fire safety education curricula or treatment protocols that may have specialized content worth considering for use with this population. It would be important to learn more about the routine or everyday services that are available in local community programs in order to better understand the nature of usual care in this area and the ways in which routine services can be enhanced. Such advances are certainly needed if the high prevalence, severity, and recurrence of this troublesome behavior are to be effectively reduced.

One recent development related to evaluation is the construction of a national database for juvenile firesetting. The Youth Firesetting Information Repository & Evaluation System is a voluntary registry that collects nonconfidential data on incidents of fire caused by children and adolescents. The data collection system provides a free summary of the statistics they receive online from fire education programs across multiple disciplines. Continued contributions from its members may produce novel assessment tools and future improvements in community-based treatment.

Fiscal Sustainability

Revenues or funding to support a practitioner or program serving the child-fire population is an important contributor to fiscal sustainability, but it is not easy to obtain or maintain. It may be a challenge to ensure that an agency has trained staff, consultants, insurance, and good relationships with local referral sources, including the fire service. And, given that there may not be a steady rate of referral, programs sometimes find it hard to keep staff fully engaged in this work. At the same time, it is important to determine a fair and equitable charge for professional services that include specialty assessment, treatment, and consultation services. Otherwise, community resources are likely to dwindle or disappear altogether. Our program services general include numerous functions related to initial case reviews (e.g., records review, note completion, review of treatment recommendations, phone consultations meetings, and treatment team meeting), administration of juvenile firesetting assessment tools (e.g., Firesetting History Screen, Firesetting Risk Interview, Fire Incident Analysis, Fire Attraction and Involvement Scale), and the development of a written summary report with service recommendations. We also provide training and conduct program evaluations to assess changes in key outcomes. We have found local service sectors more amenable to establishing focused consultation agreements to provide these services in recent years. A sample report outline is shown in Practitioner Form 15.1.

Conclusion

There are now numerous resources to facilitate intervention with children and adolescents who misuse fire and their families. Many of these resources are discussed in this book, but there is clearly a wealth of new information for professionals from numerous disciplines who work in diverse settings ranging from large multiservice programs to independent practices with a single provider. The materials developed for this book were designed to support effective service delivery in any setting and encourage the kinds of comprehensive skills likely to ensure long-term program viability. We recognize that effective management and prevention of juvenile firesetting will no doubt require continued advances in scientific knowledge and program design, the integration of complimentary educational and clinical guidelines, and developments in policy and practice that support multidisciplinary collaboration, among other topics. However, none of this innovation would have any personal or societal impact without the dedicated efforts of a cadre of energetic community practitioners and programs. We truly hope that the content of this book both supports and extends all of these programmatic efforts to promote community fire safety, child/youth skill development, and positive parent–child and family relationships.

Appendix A

Practitioner Forms

Practitioner Form 3.1

Fire History Screen

To be completed by parent or guardian:

*Name:*____________________________ *Date:*__________________

1. Does your child like fire? Yes No
2. Does your child play with matches? Yes No
3. Has your child ever burned something or set anything on fire? Yes No
4. About when did this happen?______________________________
5. What was burned? ______________________________________
6. How many times has your child burned something or set a fire? ______________________
7. How serious were the damages?

   1\. Very Minor 2. Minor 3. Moderate 4. Severe 5. Very Severe

   ____

8. Did the firefighters (fire truck) come to put out the fire? ____________________________

Questions to ask children and adolescents about misuse of fire:

9. Do you like fire? Yes No
10. Do you play with matches/lighters? Yes No
11. Have you ever burned something or set anything on fire? Yes No

Ask the next questions only if the child or adolescent indicates that he or she has burned something or set something on fire:

11. When did this happen? ______________________________________
12. What was burned? ______________________________________
13. How many times have you burned something or set a fire? ____________________________

Revised from original source: Kolko & Kazdin (1988).

Practitioner Form 3.2

Fire Incident Analysis (FIA-P)

Parent Version

NAME ______________________________________ DATE __ / __ / __

STAFF ____

To the Interviewer

The parent you will be interviewing has a child who apparently participated in a firesetting/burning incident. The questions in this interview are designed to elicit information about the fire(s), his/her perceptions of the incident, and any consequences following the act. Please use the questions as a probe. You may also want to use probing statements (e.g., "please tell me more," "think for a minute, then tell me," "can you think of anything else") if the parent is reluctant to answer or says that he/she cannot tell you. Use the lines provided to fill in the parent's responses to these questions.

To the Parent

When you talked with someone before, you told them that your child had been involved in setting ___ (number) fire(s). We want to know some more about that/those fire(s). I'm going to ask you some questions about your child and the fire(s). Please try to think hard and remember as much as you can about the fire(s). So that I won't forget what you're telling me, I will write your answers on this paper. Don't forget, this information is only between you and me. [If necessary, inquire about each fire, beginning with the most recent one].

COMMENTS ____

____

____

____

Scoring/Coding

The responses for each item in this instrument are stated along different types of scales; some are Likert-scales (intervals), and some are nominal categories for descriptive/classification purposes. Responses to those items collapsed to form dichotomies as reported in the reference are marked with a" 1" (items 5–7). Please note that there are 3 Motive scores, computed by adding together certain items, as follows: Curiosity (items 25–27), Attention (items 28 & 29), and Anger/revenge (items 30–33). There are also 3 scores to reflect the kinds of behavioral problems the child exhibited just prior to the fire, as follows: Aggression (items 10–14), Rule violations (items 15 & 16), and Depression/withdrawal (items 17–20). The remaining items (21–23) are examined individually.

Source: Kolko, D. J., & Kazdin, A. E. (1991). Motives of childhood firesetting; Firesetting characteristics and psychological correlates. *Journal of Child Psychology and Psychiatry, 32*, 535–550.

Instructions:

I would like to ask you some questions about the *most severe incident of firesetting or burning* for which your child was responsible. That is the fire that involved the (give brief description). **OK, thinking about that fire:**

1. **How many months ago did this fire (burning) take place?**____________________

____

____

____

2. **Briefly, describe the incident and what happened:**____________________

____

____

____

3. **What did he/she use to set the fire?**

1 = match
2 = lighter
3 = sparks/stove
4 = gasoline
5 = candles
6 = other
7 = no answer/don't know

4. **How did he/she get the materials?**

    1 = found it (__in/__out of) house

    2 = given by friends

    3 = went out of his/her way to get them

    4 = other (specify) ____________________

5. **Where did the fire take place (in what setting?** [Score "1" for all items but 9 and 10]

6. [If residence] **In what area or part of the home was the damage?**

1 = a suspect's bedroom	7 = living room, family room, den, etc.
2 = a parent of a suspect's bedroom	8 = basement or attic
3 = a sibling of a suspect's bedroom	9 = porch
4 = other bedroom (specify) __________	
5 = kitchen	10 = garage
6 = bathroom	

7. **What specifically was burned/damaged (what materials/objects)?**

1 = paper, tissue, cardboard	8 = firecrackers
2 = bedding, bed	9 = furniture
3 = clothing	10 = other (specify) __________
4 = toys, child's personal property	11 = others' personal property __________
5 = candle	12 = walls, floors, structure
6 = leaves, grass, trash	13 = cars/vans
7 = flammable liquid	

8. **Was anyone with him/her during this incident?** Don't know NO YES

9. [If yes] **Who?**

1 = brother	5 = friends
2 = sister	6 = other(s)
3 = siblings	7 = no answer/don't know
4 = other relative	8 = non-applicable

Further Instructions:

For the next set of questions, you will be able to choose from one of five answers. The choices are on a scale of 1 to 5, where "1" means no problem at all, and a "5" means a severe problem. [Review the scale].

How much of a problem was each of the following behaviors in the two weeks before this incident took place?

	Don't Know	Not at All		Some what		Very Much
10. not listening or minding (i.e., being stubborn)	?	1	2	3	4	5
11. tantruming or crying (whining)	?	1	2	3	4	5
12. being aggressive (i.e., hitting, kicking, or hurting others)	?	1	2	3	4	5
13. being cruel to animals	?	1	2	3	4	5
14. being angry, mad, or upset	?	1	2	3	4	5
15. skipping school	?	1	2	3	4	5
16. staying out late	?	1	2	3	4	5
17. feeling sad or blue	?	1	2	3	4	5
18. being worried/anxious	?	1	2	3	4	5
19. wetting the bed or his/her clothes	?	1	2	3	4	5
20. isolating himself/herself or withdrawing from others	?	1	2	3	4	5
21. having problems in school	?	1	2	3	4	5
22. stealing or breaking things	?	1	2	3	4	5
23. Other (specify)	?	1	2	3	4	5
______________________	?	1	2	3	4	5
______________________	?	1	2	3	4	5
______________________	?	1	2	3	4	5
______________________	?	1	2	3	4	5
______________________	?	1	2	3	4	5

24. **Briefly, why do you think this incident took place?**

1 = curiosity

2 = boredom/something to do

3 = peer pressure, influenced by friends, siblings

4 = for attention, manipulation

5 = angry, response to punishment

6 = fascinated with fire (e.g., watching, sensory, colors, shapes, thrills)

7 = likes to play with fire, burn things

8 = other (specify) ______________

9 = no answer, don't know

Now, we'd like you to indicate how true or accurate each of the following questions are in describing this incident. Here a 1 means *not at all true or accurate* and a 5 means *very much true or accurate*.

		Don't Know	Not at All		Some what		Very Much
25.	Was he/she curious about or experimenting with fire?	?	1	2	3	4	5
26.	Was this something he/she did for fun or that he/she liked?	?	1	2	3	4	5
27.	Was this a way to increase his/her feelings of power or self-confidence?	?	1	2	3	4	5
28.	Did he/she set the fire after being told not to play with matches/fire?	?	1	2	3	4	5
29.	Was he/she trying to get others to help him/her?	?	1	2	3	4	5
30.	Was he/she trying to do harm or destroy something?	?	1	2	3	4	5
31.	Was this started after he/she became angry at another person?	?	1	2	3	4	5
32.	Did this follow any family arguments or problems?	?	1	2	3	4	5
33.	Did this follow any school problems?	?	1	2	3	4	5
34.	Did this follow any problems with his/her friends?	?	1	2	3	4	5
34a.	Was this incident an accident?	?	1	2	3	4	5

We would like to ask you a few questions about the effects of this incident; that is, what happened after it took place?

35. **Child Responses**

a) Did he/she panic when it happened?	Don't Know	No	Yes
b) Did he/she try to get help?	Don't Know	No	Yes
c) Did the child deny responsibility?	Don't Know	No	Yes
d) Did he/she watch any part of the fire?	Don't Know	No	Yes
e) Did he/she hide after the incident?	Don't Know	No	Yes

36. **Family Circumstances**

a) Did the family pay special attention to him/her after this incident?	Don't Know	No	Yes
b) Was he/she talked to, counselled, or treated by someone outside of the family?	Don't Know	No	Yes

c) [If Yes] How so? By whom? What happened?

1 = Fire & Rescue Service
2 = Neighbour/friend
3 = Relative
4 = Other non-mental health professional
5 = Mental health professional/counsellor or therapist
6 = Other ______________________

37. **Discipline**

a) Was there any discipline after this incident? Don't Know No Yes

b) [If Yes] What?

1 = Verbal punishment (yelled at, screamed at, etc.)
2 = Physical punishment
3 = Isolation (sent to room, sent to corner)
4 = Grounded, withdrawal of privileges
5 = Discussion of behaviour
6 = Taken to professionals
7 = Other ______________
8 = No Answer

38. **Financial**

What was the general value (cost) of the damages? ______________________

39. **Medical**

a) Was anyone hurt or injured? Don't Know No Yes

b) [If Yes] Who was injured? ______________________

c) [If Yes] What was (were) the injury (injuries)?

1 = Minor burns/scrapes 3 = Serious burns

2 = Moderate 4 = Other ______________

d) Did anyone die in the incident? Don't Know No Yes

e) [If Yes] Who? ______________________

40. **Legal**

a) Did this incident lead to any juvenile record? Don't Know No Yes

b) [If Yes] What kind (charge)? ______________________

____

c) Did the incident lead to his/her being removed from the home? Don't Know No Yes

d) Were there any other legal effects? Don't know No Yes

e) [If Yes] What were they? ______________________

____

____

41. **Social**

a) Did his/her friends avoid or reject him/her after this (i.e., saw him/her as a trouble-maker)? Don't Know No Yes

b) Did his/her friends begin to accept him more or spend more time with him/her (i.e., saw him/her as a big shot)? Don't Know No Yes

c) [If Yes] How so? By whom? What happened?

1 = Fire & Rescue Service 4 = Other non-mental health professional

2 = Neighbour/friend 5 = Mental health professional/counsellor or therapist

3 = Relative 6 = Other ______________

42. **Other Consequences**

a) Did anything else happen due to this incident (any other consequences)? Don't Know No Yes

b) [If Yes] What? Don't Know No Yes

1 = Hid or removed firesetting materials

2 = Further evaluation/inquiry

3 = Alerted, worried, monitored, scared

4 = Bragging, increased intrigue with fire

5 = Restriction/lost privileges

6 = Education/sent for treatment

7 = Other ________________

42. Is there anything else about this incident that we haven't asked about that would be important?

____

____

____

____

____

THANK YOU FOR YOUR INFORMATION. THIS INFORMATION WILL ENABLE US TO LEARN MORE ABOUT CHILDREN AND THEIR USE OF FIRE.

Practitioner Form 3.3

Fire Incident Analysis (FIA-C)

Child Version

NAME ______________________________ DATE___ / ___ / ___

STAFF ______________________________

To the Interviewer

The child you will be interviewing has apparently participated in a firesetting/burning incident. The questions in this interview are designed to elicit information about the fire(s), his/her perceptions of the incident, and any consequences following the act. Please use the questions as a probe. You may also want to use probing statements (e.g., "please tell me more," "think for a minute, then tell me," "can you think of anything else") if the child is reluctant to answer or says that he/she cannot tell you. Use the lines provided to fill in the child's responses to these questions.

To the Child

When you talked with someone before, you told them that you had been involved in setting ____ (number) fire(s). We want to know some more about that/those fire(s). I'm going to ask you some questions about the fire(s). Please try to think hard and remember about the fire(s). So that I won't forget what you're telling me, I will write your answers on this paper. Don't forget, this information is only between you and me. [If necessary, inquire about each fire, beginning with the most recent one].

COMMENTS ______________________________

____

____

____

Scoring/Coding

The responses to each item in this instrument are not stated along interval scales (e.g., Likert-scale), but, rather, reflect nominal categories for descriptive/classification purposes. Responses to these items were collapsed to form dichotomies in the direction of increased problem severity, as reported in the article listed below (i.e., 0 vs. 1 in parenthesis following each response). Please note that items 1, 2, and 17 were meant to elicit information for use in surveying other items in the instrument and, thus, were not coded in this study.

Source: Kolko, D. J., & Kazdin, A. E. (1993). Children's descriptions of their firesetting incidents: Characteristics and relationship to recidivism. Journal of the American Academy of Child and Adolescent Psychiatry, 33, 114–122.

1. **When did this fire take place? About how many months ago?** ____________________

____

____

____

2. **What happened? [Use this information to help score subsequent questions]:**

____

____

____

3. **What did you use to set the fire?**

 1 = match (1)

 2 = lighter (0)

 3 = sparks/stove (0)

 4 = other (e.g., gasoline, candles) (0): ____________________

 98 = no answer/don't know

4. **How did you get the ________ to start the fire?**

 1 = found it in house (0)

 2 = found it outside house (0)

 3 = given by someone (0)

 4 = went out of way to get it, e.g. bought it, opened cabinets (1)

 5 = other: ____________________

 98 = no answer/don't know

5. **What did you burn—what things got burned/damaged?**

1 = paper, tissue, cardboard (0)
2 = leaves, brush/grass, wood, trash (0)
3 = toys/child's personal property (0)
4 = clothing (0)
5 = sheets/towels (0)
6 = other's personal property (1)
7 = furniture (1)
8 = walls, floors, structure (1)
9 = cars/vans (1)
10 = self (1)
11 = other people (1)
12 = other: ______________________
98 = no answer/don't know

6. **Where did the fire take place? Where did you do this?**

1 = own bedroom (0)
2 = a parent of a suspect's bedroom (0)
3 = a sibling of a suspect's bedroom (0)
4 = other bedroom (0)
5 = kitchen (0)
6 = bathroom (0)
7 = living room, family room, den, etc. (0)
8 = basement or attic (0)
9 = outside porch (1)
10 = garage (1)
11 = the child's yard (1)
12 = neighbour's yard (1)
13 = vacant lot or street (1)
14 = occupied single family dwelling (1)
15 = mercantile (1)
16 = shed or other building (1)
17 = occupied multiple family dwelling (1)
18 = school (1)
19 = church (1)
20 = car or truck (1)
21 = residential treatment facility (1)
22 = other: ______________________
98 = no answer/don't know
99 = non-applicable

7. **Was anyone with you when you did this?** 0 = NO 1 = YES N/A

8. [If yes] **Who?**

1 = brother
2 = sister
3 = siblings
4 = other relative
5 = friends
6 = other(s): ______________________
7 = no answer/don't know
8 = non-applicable

9. **Why did you do this—what was the main reason you did this?**

1 = curiosity/like fire (0)

2 = experimentation/looks/movements, etc. (0)

3 = fun/playing around/boredom, just wanted to (0)

4 = to be warm, other non-interpersonal functional purpose (1)

5 = for some functional purpose (ie manipulation, influence) (1)

6 = mad/emotional release (ie response to some event) (1)

7 = encouraged by friends (1)

8 = other: ______________________

98 = no answer/don't know

10. **Was there another reason why you did this?**

0 = no

1 = yes

98 = no answer/don't know

11. [If yes] **What?**

1 = curiosity/experimentation (0)

2 = for some functional purpose (1)

3 = encouraged by friends (1)

4 = mad/emotional release (1)

5 = revenge (1)

6 = other: ______________________

98 = no answer/don't know

99 = non-applicable

12. **How long did you think about doing this?**

1 = no forethought; decided right then (0)

2 = 1–30 minutes (1)

3 = sometime that day/within 24 hours (1)

4 = between 1 and 2 days (1)

5 = more than 2 days (1)

98 = no answer/don't know

13. **Did you plan out how you were going to do this?**

0 = no

1 = yes

98 = no answer/don't know

14. [If yes] **How did you plan it out?**

1 = stored up/saved materials (0)

2 = considered most efficient/unique method (0)

3 = considered safest/most concealed method (i.e., to avoid being caught) (0)

4 = stored up materials & considered efficiency or safety method; any combo of above (1)

5 = other

98 = no answer/don't know

99 = non-applicable

15. **What did your parent(s)/guardians do after you did this?**

1 = they never knew (0)

2 = nothing (1)

3 = reprimand, scolding, got mad or other emotional reaction (verbal punishment) (0)

4 = education, discussion of dangers (0)

5 = sent to room (0)

6 = grounding (0)

7 = physical punishment (0)

8 = took to fire department for education (0)

9 = other

98 = no answer/don't know

99 = not applicable

16. **Do you have any brothers or sisters living at home?**

0 = no

1 = yes

98 = no answer/don't know

17. [If yes] **What did they do after you did this? (select primary reaction)**

1 = they never knew (0)

2 = nothing (1)

3 = reprimand, scolding, got mad (0)

4 = approval/positive (1)

5 = mixed reactions (some positive/some negative) (1)

6 = other

98 = no answer/don't know

99 = not applicable

18. [If yes] **What did your friends do after you did this? (select primary reaction)**

1 = they never knew (0)

2 = nothing (1)

3 = reprimand, scolding, got mad (0)

4 = approval/positive (1)

5 = mixed reactions (some positive/some negative) (1)

6 = other

98 = no answer/don't know

99 = not applicable

19. **Do you think you will do this again?**

0 = no (code "99" for #20, then ask #21)

1 = yes (ask #20, then code "99" for # 21)

98 = no answer/don't know

20. [If yes] **Why?**

1 = fun (1)

2 = boredom/something to do (1)

3 = no reason (0)

98 = no answer/don't know

99 = not applicable

Practitioner Form 3.4

Firesetting Risk Interview (FRI)

(Parent)

Instructions for Administration and Scoring

The FRI was designed to evaluate several dimensions believed to be associated with firesetting, based on the clinical and empirical literature. For purposes of clarification, there are 86 items in the FRI (not 92, as was implied in the article). The item pool began with 92 (not 99) items, six of which were deleted for psychometric reasons.

Due to the inclusion of diverse items, the various dimensions or sections incorporate alternative response choices. This makes the interview somewhat complicated to complete. For this reason, the measure has been administered in interview format. The dimensions reflect fire-specific and non-fire-specific content, as reflected in the following:

<u>Specific to Fire</u>

Curiosity about Fire (A)
Knowledge of Fire Safety (B)
Fire Skill/Competence (C)
Complaints/Concern about Fire Behavior (G)
Exposure to Peer/Family Models (H)
Involvement in Fire-Related Activities (I)
Parental Fire Awareness/Preparation (J)
Early Experiences with Fire (K)

<u>Not Specific to Fire</u>

Expression of Positive Behaviors (D)
Expression of Negative Behaviors (E)
Exposure to Supervision/Discipline (F)
Frequency of Mild Punishment (L)
Efficacy of Mild Punishment (M)
Frequency of Harsh Punishment (N)
Efficacy of Harsh Punishment (O)

The FRI is scored simply by summing the items in (and deriving a separate score for) each of the 15 dimensions labeled by letters. Because the specific items associated with the final four dimensions may be difficult to determine, the items in each dimension are described here: L (sum items 71, 73, 75, 77, and 79), M (72, 74, 76, 78, 80), N (81, 83, 85), and O (82, 84, 86).

The measure is described in the following source: Kolko, D. J., & Kazdin, A. E. (1989). Assessment of dimensions of childhood firesetting among patients and nonpatients: The Firesetting Risk Interview. *Journal of Abnormal Child Psychology 17,* 157–176.

Firesetting Risk Interview (FRI)

(Parent)

ID: ____________________________ Your child's name: ______________________________

Your name: ____________________ Your relationship to child: _______________________

Today's Date: _____ / ______ / ______ Administered by: ______________________________

We are asking you to provide information about your child's exposure to and involvement with fire. You will also be asked other questions about fire and your child's behavior in general. Most of the questions can be answered by selecting a number from 1 to 5, although some questions ask you to use your own words.

A. Curiosity About Fire

	Not at All		Some what		Very Much
1. How curious is he/she about fire?	1	2	3	4	5
2. How much does he/she want to play with fire?	1	2	3	4	5
3. How much does he/she think that fire is special or magical?	1	2	3	4	5
4. How much does he/she get excited or fascinated when fires or fire-related topics are mentioned in everyday conversation?	1	2	3	4	5
5. How much does he/she like to talk about fire?	1	2	3	4	5
6. How much does he/she want to visit exhibits or watch movies about fire, or to actually watch a real fire?	1	2	3	4	5
7. How much does he/she read and attempt to learn about fire and its uses?	1	2	3	4	5

B. Knowledge of Fire Safety

	Not at All		Some what		Very Much
8. To what extent does your child understand his/her own behavior, in general?	1	2	3	4	5
9. To what extent does he/she know different facts about fires or firefighters?	1	2	3	4	5

	Not at All		Some what		Very Much
10. To what extent does he/she understand why playing with fire is dangerous?	1	2	3	4	5
11. To what extent does he/she know what things will burn and what things won't?	1	2	3	4	5
12. To what extent does he/she know how to use matches or lighters correctly?	1	2	3	4	5

C. Fire Skill/Competence

	Not at All		Some what		Very Much
13. To what extent does he/she know what to do if something catches fire suddenly?	1	2	3	4	5
14. To what extent has he/she been taught to use matches or lighters correctly?	1	2	3	4	5
15. To what extent does he/she play safely when alone or with others?	1	2	3	4	5
16. To what extent is he/she able to light a fire and put it out correctly?	1	2	3	4	5
17. To what extent is he/she allowed to use matches or lighters at home?	1	2	3	4	5

D. Expression of Positive Behaviors

How often does he/she express himself/herself in each of the following ways?: [positive behaviors]

	Not at All		Some what		Very Much
18. By touching or using some form of pleasant physical contact	1	2	3	4	5
19. By complimenting or praising others	1	2	3	4	5
20. By laughing or using humor/jokes	1	2	3	4	5
21. By providing attention to others	1	2	3	4	5
22. By making pleasant conversation	1	2	3	4	5

E. Expression of Negative Behaviors

How often does he/she express himself/herself in each of the following ways?: [negative behaviors]

	Not at All		Some what		Very Much
23. By hitting or hurting others	1	2	3	4	5
24. By criticizing or disapproving of others	1	2	3	4	5
25. By giving orders or making threats	1	2	3	4	5
26. By ignoring others or not doing anything at all	1	2	3	4	5
27. By yelling or screaming at others	1	2	3	4	5
28. By being stubborn or not minding others (not following instructions)	1	2	3	4	5
29. By destroying property	1	2	3	4	5
30. By crying or whining	1	2	3	4	5
31. By being cruel to animals	1	2	3	4	5
32. By threatening to hurt or actually doing something to hurt himself/herself	1	2	3	4	5

F. Exposure to Supervision/Discipline

	Not at All		Some what		Very Much
33. How often is he/she unsupervised at home by you or another adult, in general?	1	2	3	4	5
34. How often do you supervise him/her at home?	1	2	3	4	5
35. How often is he/she supervised by an adult when he/she is with friends?	1	2	3	4	5
36. How often is your child disciplined at home?	1	2	3	4	5
37. How often does your child receive attention from family members?	1	2	3	4	5
38. How often is your child disciplined by others outside of the home (adults in the community, teachers, etc.)	1	2	3	4	5

G. Complaint/Concern About Fire Behavior

	Not at All		Some what		Very Much
39. How often do you receive complaints about his/her behavior, in general, from others in the community?	1	2	3	4	5
40. To what extent do you receive complaints about his/her play with fire from others in the community?	1	2	3	4	5
41. How often do you worry about him/her playing with fire when he/she is left unattended?	1	2	3	4	5

H. Exposure to Peer/Family Models

	Not at All		Available/ Not Easy to Get to		Available/ Easy to get to
42. How available are matches, lighters, or other fire-starting materials at his/her school or in his/her friends' homes?	1	2	3	4	5
43. How available are matches, lighters, or other fire-starting materials in or around your home?	1	2	3	4	5

	Not at All		Some of the Time		Almost Always
44. How often is he/she in the presence of friends who smoke anywhere outside your home (e.g., school, friends' homes)?	1	2	3	4	5
45. How often is there cigarette or pipe smoking in your home?	1	2	3	4	5

	None	One	Two	Three	Four/ More
46. How many times have other family members been burned or hurt because of a fire in the last year?	1	2	3	4	5
47. How many people who live at home (including yourself) smoke cigarettes or pipes?	1	2	3	4	5

	None	One	Two	Three	Four/ More
48. How many family members have an interest or fascination with fire?	1	2	3	4	5
49. How many family members has he/she observed playing with matches or lighting fires in the last year?	1	2	3	4	5
50. How many other persons in your neighborhood have been burned or hurt because of a fire in the last year?	1	2	3	4	5
51. How many times has he/she ever been burned or hurt because of a fire in the last year?	1	2	3	4	5
52. How many times have other family members been burned or hurt because of a fire in the last year?	1	2	3	4	5
53. How many of his/her friends smoke or experiment with smoking?	1	2	3	4	5
54. How many fires have there been in your neighborhood in the last year?	1	2	3	4	5

I. Involvement in Fire-Related Activities

	None	One	Two	Three	Four/ More
55. How many times has your child ever hidden matches, lighters, or other fire-starting materials?	1	2	3	4	5
56. How many times has your child left burn marks on things in your home?	1	2	3	4	5
57. How many times has anyone such as school officials, the police, or your neighbors told someone in your family about your child's playing with fire?	1	2	3	4	5

J. Parental Fire Awareness/Preparation

Please answer No (0) or Yes (1) for these questions.

	No	Yes
58. Do you usually give instructions about fire to the babysitter or others who take care of your children?	0	1
59. Do you know the phone number for the neighborhood fire department?	0	1
60. Do you have access to a fire extinguisher?	0	1
61. Is there a smoke alarm in your home?	0	1
62. Have you ever received any formal fire education or training?	0	1
63. Have you ever received any guidance or general information about children playing with fires?	0	1
64. Have you ever told your child why it is bad to play with fire?	0	1
65. Have you ever practiced fire escape drills with your child?	0	1

K. Early Experiences with Fire

	No	Yes
66. Were there any smokers living in your home more than 1 year ago?	0	1
67. Did any members of your family play with amtches or lighters, or light fires more than 1 year ago?	0	1
68. Was your child exposed to any neighborhood fires, or to other people who played with fire more than 1 year ago?	0	1
69. More than 1 year ago, did you child ever play with matches/lighters or fire?	0	1
70. Did your child ever show any special interest in fire more than 1 year ago?	0	1

L–O. Frequency and effectiveness of Punishment

For each of the different methods of disciplining this child listed below, please indicate: a) how often each is used, and b) how effective/helpful it usually is.

Note: If answer to "How often" is "1", leave "How effective" blank.

		L. How Often:						**M. How Effective:**				
		Not at All		Some of the Time		Almost Always		Not at All		Some of the Time		Almost Always
Isolation or some form of quiet time	71.	1	2	3	4	5	72.	1	2	3	4	5
Taking away things or privileges	73.	1	2	3	4	5	74.	1	2	3	4	5
Extra work or chores to do	75.	1	2	3	4	5	76.	1	2	3	4	5
Extra support, attention, or affection	77.	1	2	3	4	5	78.	1	2	3	4	5
Discussion and review of behavior	79.	1	2	3	4	5	80.	1	2	3	4	5

		N. How Often:						**O. How Effective:**				
		Not at All		Some of the Time		Almost Always		Not at All		Some of the Time		Almost Always
Spanking, slapping, or some other form of physical punishment	81.	1	2	3	4	5	82.	1	2	3	4	5
Reprimands, yelling, or screaming	83.	1	2	3	4	5	84.	1	2	3	4	5
Threatening or screaming	85.	1	2	3	4	5	86.	1	2	3	4	5

Practitioner Form 3.5

Children's Firesetting Interview (CFI)

Child's Name	Interviewer	Date:	__ / __ / ___
____________________	______________________________	ID:	__________

Instructions to the Interviewer

Administration: This semistructured interview is designed to elicit information from the child regarding several dimensions which may help up to better describe and predict firesetting behavior. The interview contains *two different formats for recording the child's responses*. The more common format (multiple-choice) involves recording the child's response that matches one of the available answers included on this form, such as a number from "1 to 5" or a "Yes or No". To help the child remember what each available answer choice means, a card that lists each selection should be explained and given to the child before questioning begins.

The other format (fill-in) involves simply writing down the child's answer verbatim, such as an explanation or a number. It should be obvious when each format is employed.

If the child says he/she doesn't know an answer, say "Are you sure?" If the child fails to respond, use one or two prompts to elicit a response (e.g., "What would you say?" "Think about it for a second and then tell me."). If these prompts fail, simply fill in a "DK" (don't know) in the space provided.

Scoring: To score the CFI, simply sum the scores for the items in each domain (A, B, C, etc.). Use this information to identify domains that suggest low levels of desirable behaviors (C, D, F) or high levels of undesirable behaviors (A, B, E, etc.).

Instructions to the Child

We are interested in what children, like you, know about fire. Please listen carefully to my questions and try to answer each question as best you can. For some of the questions, I'll show you a card from which you can choose the best answer. The answer could be: 1) not at all true, 2) very little, 3) somewhat, 4) a lot, or 5) very much true. Here, you pick only one of these answers or choices.

For the other questions, I just want you to answer the question in your own words. I'll write down your answers as we go through the questionnaire.

Certain items were based on questions contained in the Child Questionnaire published by the Federal Emergency Management Agency (1979), Washington, D.C. Item #35 is based on the work of Jones, R. T., Kazdin, A. E., & Haney, J. (1981), *Journal of Applied Behavior Analysis, 14,* 249–260.

*Note: items 31 and 33 were revised in this version (1/1/2016). Original: 10/96, 1/90

Children's Firesetting Interview (CFI)

Child's Name ____________________ Interviewer ____________________ Date: ___ / ___ / ___

ID: ____________

Scale for questions 1–8: (Give card to child): 1 = not at all; 2 = very little; 3 = somewhat; 4 = a lot; 5 = very much

A. Curiosity about (or Attraction to) Fire

1. How curious are you about fire (i.e., want to know more about it)? _______ 1
2. How much do you think about fire? _______ 2
3. How much do you want to play with fire? _______ 3
4. How special or magical is fire to you? _______ 4
5. How excited or interested do you get when people talk about fires? _______ 5
6. How much do you like to visit exhibits or movies about fires, or to watch a real fire? _______ 6
7. How much do you like to read about and learn about fire, and the right way to use it? _______ 7
8. How much do you like to talk about fire, rather than other things? _______ 8
9. What do you like most about fire?[1] _______ 9
10. When you think about fire, what do you think about?[1] _______ 10

B. Involvement in Fire-related Activities

11. Did you ever pull a fire alarm when there really wasn't any fire or smoke around? _______ 11

    1 = NO

    YES (If yes: How many times?)

    2 = One

    3 = Two

    4 = Three to Five

    5 = Six or More

[1] "Other responses should be assigned a score for that item with which they are most closely associated; otherwise, assign a middle score of 3.

12. Did you ever hide matches, lighters, or other fire-starting materials? ______ 12

1 = NO

YES (If yes: How many times?)

2 = One

3 = Two

4 = Three to Five

5 = Six or more

13. Did you ever leave burn marks on things in your home? ______ 13

1 = NO

YES (If yes: How many times?)

2 = One

3 = Two

4 = Three to Five

5 = Six or more

14. Did anyone, like someone from the school, the police, or your neighbors, ever tell someone in your family about your playing with fire? ______ 14

1 = NO

YES (If yes: How many times?)

2 = One

3 = Two

4 = Three to Five

5 = Six or more

C. Knowledge About Combustibles

Now, I'm going to mention a lot of different things. I want you to tell me if it will or will not burn when you touch it with a lighted match. So, answer YES (it will) or NO (it won't) for each item. Guess if you have to. (Code number corresponding to answer):

	0	1			
15. Chalk that you would use to write on a blackboard	Y	N	DK	______	15
16. Aluminum cans, like a pop/soda/Coke can				______	
17. Pieces of wood				______	
18. Glass, like a window				______	
19. Clothes, like a shirt or pair of socks				______	
20. Bricks that are used to make houses or buildings				______	
21. Steel or metal, like a matchbox car or the parts of a large building				______	
22. Gasoline used to make cars go				______	
23. Orange juice or apple juice				______	
24. Skin, like on your hand or face				______	
25. Chocolate milk				______	
26. Baby lotion, like what mom or dad would put on a small child				______	
27. Shampoo for your hair				______	
28. Bleach that you might use to clean clothes				______	
29. A couch, like the kind you sit on				______	

D. Fire Safety Skills

30. Are there any dangers to playing with fire? ______________ 30

1 = No/Don't Know

2 = General problem (get in trouble, people get mad at you)

3 = Burn things (damage)

4 = Burn self/others (hurt, killed)

5 = Burn things & people

31.* If you needed to use a match to light a few pieces of wood to make a campfire on a beach, what steps would you follow—what would you do first?; what would you do next? (score 1 point for each item listed below and enter total score at right) _______________ 31

Total # of correct steps [code "0" if child does not know or none are correct]

☐ 1 Dig a hole or set up a ring/boundary area around the wood to contain the fire (e.g., make pit, use sand to make boundary)

☐ 2 Put some pieces of wood on top of the fire starter

☐ 3 Put some wood on top of fire starter

☐ 4 Strike the match and light the fire starter (i.e., not throwing match on wood pile)

☐ 5 Have water or extinguisher (or related material) nearby in case of emergency

32. How would you put out this campfire on the beach? _______________ 32

1 = Questionable behavior (i.e., use a poker, stamp out); Don't know

2 = Let it burn, don't add more wood

3 = Smother, cover, beat

4 = Water/Chemical substances (e.g., baking soda)

5 = Use extinguisher

33.* Now, if an adult needed to light a candle using a match, what steps would the adult need to follow in order to do this safely? Let's assume the adult has a pack of matches, like this one—now, don't worry, this won't really light so we can be safe, but you can act as if they are real [GIVE FAKE MATCH PACK] (Score 1 point for each correct answer—if child doesn't know, score 0) _______________ 33

☐ 1 Removes match from the pack

☐ 2 Closes cover

☐ 3 Strikes away from body

☐ 4 Places fingers away from match head

☐ 5 Strikes match on its side

34. What would you do if your clothes caught on fire? (select primary answer if more than 1 are given) ______________ 34

1 = Don't know/nothing/run/run to water

2 = Call for help (tell grown up)/helpless behavior (cry)

3 = Throw water on them

4 = Smother/behavior involving taking off clothes

5 = Stop, drop, and roll

35. Imagine that you're sitting on your bed when a fire starts in your house. You start to cough and your eyes are burning, and you can't leave through a window. Tell me everything you would do. (Score one point for each item listed below.) ______________ 35 Total # of correct items [code "0" if none or D/K]

- ☐ 1 Slide to edge of bed
- ☐ 2 Roll out of bed
- ☐ 3 Get in a crawl position
- ☐ 4 Crawl and get the rug, blanket, etc.
- ☐ 5 Push rug in crack
- ☐ 6 Crawl to window
- ☐ 7 Open window
- ☐ 8 Yell and signal for help

36. Has your family ever discussed a plan to follow if there was a fire in the house? ______________ 36

(If Yes: what is it?)

1 = None

3 = Self-plan only/some general details (i.e., family has a plan but child doesn't know it)

5 = Clear/reasonable family plan (i.e., involves other members)

37. Pretend that you are home all alone and that you suddenly see a fire in one of the rooms. Here's a real telephone. How would you use the telephone to get help? What would you do? (Act it out; Show me, don't tell me) [This task requires the use of a real/play phone.] ____________ 37

DIALS

☐ 1 Picks up receiver

☐ 2 Places receiver to ear

☐ 3 Dials "0" or "911" (NOT a random or incomplete set of numbers)

REPORTS THE FIRE

☐ 4 "There is a fire at my house" (Summarize the remarks)

____

____

GIVES NAME

☐ 5 First name

☐ 6 Last name

Other Comments:

____

GIVES ADDRESS

☐ 7 House number

☐ 8 Street name

☐ 9 Town/Municipality (i.e., section of the city)/City

Other Comments:

____

E. Exposure to Models

38. How many of your friends smoke? _______ 38

1 = None

2 = One

3 = Two

4 = Three to five

5 = Six or more

39. How many of your friends have you seen playing with matches or lighters, or setting fires? _______ 39

1 = None

2 = One

3 = Two

4 = Three to Five

5 = Six and more

40. Are you permitted to use matches or lighters at home? What for?: _______ 40

____

1 = NO

3 = YES (restricted use/under supervision)

5 = YES (unrestricted use)

Scale for #41–46: (Give card with scale to child)

Scale: 1 = never, 2 = a little, 3 = sometimes, 4 = most of the time, 5 = almost always

I have just a few more questions. Please pick the best answer from this card (review choices before administering questions):

41. How often are you with friends who smoke? _______ 41

42. How often do you smoke or experiment with smoking? _______ 42

43. How often are you supervised by an adult when you are in your house? _______ 43

____

F. Discipline/Supervision

44. How often are you disciplined at home when you do something that you are not supposed to do—something wrong (i.e., corrected by your parents)? ____ 44

45. How often are you disciplined by other people outside of your home when you do something you are not supposed to do? ____ 45

46. How often are you supervised by an adult or an older brother/sister when you are at your friends' homes (i.e., someone who is watching over you)? ____ 46

Practitioner Form 4.1

Fire Graphing Sheet

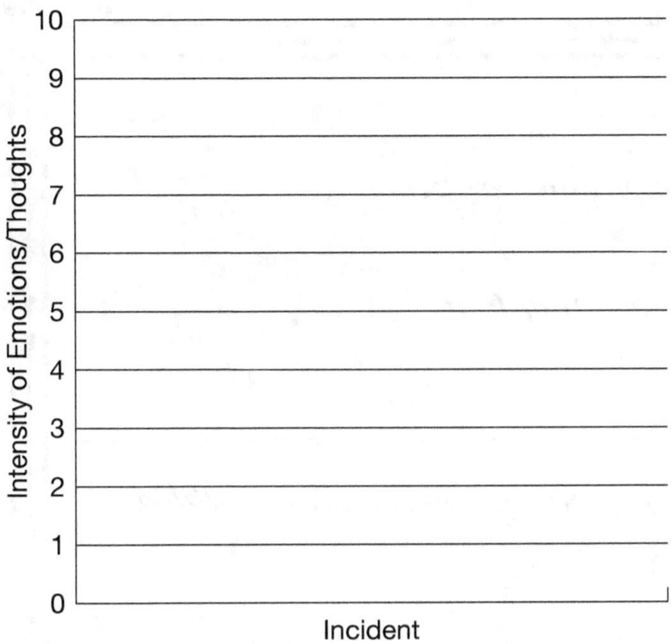

Events/Situations

List each at the bottom, from first to last.

Label each emotion or thought here:

● ______________________

× ______________________

■ ______________________

▲ ______________________

◆ ______________________

Practitioner Form 14.1

Program Completion Certificate

Program Completion Certificate

This is to certify that

____

has successfully completed a firesetting intervention, entitled

_______________________ _______________________ _______________________

Date

Practitioner Form 15.1

Sample Consultation Report with Recommendations

Child's Name ______________________________ Assessment Date: ___________

Completed by ______________________________ ID# _______

Brief Child and Family Background, and Description of Incidents

____

____

____

____

Current Impressions/Diagnosis:

____

____

____

____

Major Stressors:

____

____

Intervention/Treatment Recommendations:

____

____

Reviewed by:

__________________________ ________________

Provider Date

__________________________ ________________

Manager Date

Appendix B

Child Forms

Child Form 6.1

The Fire Triangle

Child Form 9.1

All About Feelings

- Having feelings is normal. All of us have feelings.
- Feelings are inside us. You don't see them.
- People may feel differently about the same thing.
- Some feelings are pleasant and some are unpleasant, but all feelings are ok to have and talk about.
- Feelings are important and important to learn about.
- Feelings can change, especially when we understand something better (that is, when we know more about it).
- We can tell about someone's feelings by:
    - Looking
    - Listening
    - Asking Questions

Child Form 9.2

How Do Others Feel?

1. Susan's birthday is tomorrow. As a surprise, her parents told her they would be going to her favorite beach for a weekend trip.

   She probably feels ______________________________

2. Tanika just heard that her grandma is very sick and might not live very long. She really loves her grandma very much.

   She probably feels ______________________________

3. Erin is watching a weather report on TV. She learns that there is a heavy rainstorm near the airport. Her dad has just left to drive to the airport to go out of town for work, but she cannot reach him by phone.

   She probably feels ______________________________

4. Mario's uncle has come for a visit. His uncle brings everyone in the family a gift. This time, he brings Mario's older brother a football, his younger sister a big doll, and his younger brother a videogame set. He gives Mario a deck of playing cards.

   He probably feels ______________________________

5. Jackson can't wait to go to a monster jam at the arena, since his dad told him that they would be going later that night. Jackson has come home early to get dressed and put on his favorite NASCAR shirt.

   He probably feels ______________________________

6. Matt and Talib are friends. They are building a small robot from a kit in class. Darrell walks up and says to Matt, "Are you some kind of nerdy scientist? I thought you were going to play soccer with us."

   Matt probably feels ______________________________

Child Form 9.3

How to Relax

General Muscle Relaxation

1. Take some slow, long, deep breaths.
2. Take a big breath in, tense your entire body (hold it still), count to five, and then breathe out and completely relax or let go.
3. Continue tensing and relaxing your body 5 more times.
4. Remember to take deep breaths.
5. Let your eyes drift open.
6. Pay attention to how relaxed you feel.

For a Deeper Relaxation

1. Pick a muscle (area) where you feel tension or tightness, such as your legs, buttocks, arms, or face.
2. Tense each muscle for 5 seconds, and then relax for 5 seconds, focusing on how each muscle feels.
3. Once you have done this with each muscle area, then try to relax your entire body.
4. Remember to take deep breaths as you do this exercise.
5. Repeat this sequence a second time, if you can.
6. Pay attention to how relaxed you feel.

Child Form 9.4

Different Thoughts—Same Situation

Situation: You wake up in the middle of the night to a strange noise.		
	Thoughts	Feelings
Response 1		
Response 2		

Situation: You get your first A in a class.		
	Thoughts	Feelings
Response 1		
Response 2		

Situation: You lose your favorite t-shirt.		
	Thoughts	Feelings
Response 1		
Response 2		

Child Form 10.1

A Problem is Something That …

- Makes someone have upset feelings.

- Happens mostly between people, but sometimes not.

- Needs to be solved (we have to do or say something to make the problem go away).

To summarize: *A problem gives someone unhappy or upset (unwanted) feelings and needs to be solved.*

Child Form 10.2

Identifying Problem Situations

Look at the first two pictures. What feeling could each person have?

- How can you tell?
- What problem could she be having?
- What is the goal?

Now, look at the next two pictures. What is going on here?

- What feeling could they be having?
- How can you tell?
- Is this person having a problem?

Child Form 10.3

Alternate Uses Worksheet

Instructions: Think of all the ways in which you could use these items.

Pencil: ____________________

Trash can: ____________________

Popsicle stick: ____________________

Tire: ____________________

Plastic ring from 6-pack: ____________________

Leaf: ____________________

Balloon: ____________________

Soap dish: ____________________

Coconut: ____________________

Candle: ____________________

Marble: ____________________

Brick: ____________________

Piece of wire: ____________________

Shoe lace: ____________________

Ice cube tray: ____________________

Cereal box: ____________________

Child Form 10.4

Solutions and Consequences Worksheet

Read each problem shown in the first column. Then answer the following questions:

1) What are the possible solutions?
2) What might happen next?
3) Will the solution lead to what you want to happen (without other bad things happening too)?

Problems	Possible solutions	What might happen if you do this (consequences)
Imagine that you ripped your brand new pair of jeans while playing with your friends, when your mother told you to only wear them for school.		
You borrowed your friend's new comic book and it was stolen from your book bag.		
You had a fight with your mother. Now you want to ask her if she will drive you to the mall to meet your friends.		
You forgot to throw your dirty clothes in the wash, and you have nothing clean to wear for school.		
Last week you heard some kids in the bathroom talk about stealing something from a store. This week you heard that the store had been robbed.		

Child Form 10.5

My Favorite Things Worksheet

1. My favorite activities after school are:

____

____

____

2. My favorite activities on the weekends are:

____

____

____

3. What groups are you in, like sports teams, clubs, or music lessons?

____

____

____

4. Circle yes or no

Yes	No	There are children to play with in my neighborhood.
Yes	No	There is a park I can play in close to my home.
Yes	No	I think I need more children to play with.
Yes	No	I think I need more after-school activities.

5. What are some times when you don't have enough to do, when you feel lonely or bored?

____

____

6. What are some activities you would like to do more often?

____

____

Child Form 10.6

Problem-Solving to Prevent Fires

Step 1: What is a problem or situation where you could get involved with fire?

____

____

Step 2: What would be your goal in that situation?

____

Steps 3 and 4: What are some possible solutions to that problem? What might be the consequence of each solution?

Solution	Consequence

Step 5: Which is the best solution and why?

____

Step 6: What steps would you need to take to make that solution happen?

____

____

Step 7: How would you evaluate the outcome? What things would you think about to see if it was a good or bad solution?

____

Child Form 10.7

Problem-Solving Worksheet

Example: Your dad is watching the news but you want to watch something else.

Your Problem:	You want to watch TV but someone else already is.
Your Feeling:	Disappointed, Bored
Your Goal:	Watch something better.

Here are some practice situations for you to read and think about as if they were happening to you. Say what the problem might be, how you might feel, and what your goals are.

1. **Your sister has the TV on too loud, making it hard for you to do your homework that is due tomorrow.**

   Your problem: ______________________________

   Your feeling: ______________________________

   Your goal: ______________________________

2. **A friend in your math class asks if he can cheat off of your test.**

   Your problem: ______________________________

   Your feeling: ______________________________

   Your goal: ______________________________

3. **You borrowed your brother's cell phone without asking him and you dropped it. Now, it doesn't work.**

   Your problem: ______________________________

   Your feeling: ______________________________

   Your goal: ______________________________

Now think of 2 problems that you had in the past or one that you are having right now. Write it down and then write down your feelings and goals.

1. Your problem: ____________________

   Your feeling: ____________________

   Your goal: ____________________

2. Your problem: ____________________

   Your feeling: ____________________

   Your goal: ____________________

Child Form 10.8

Practice at Home Worksheet

Next time, we will discuss problems that were difficult for you to solve this week. We will also talk about how you handled them. They can be any kind you want.

Examples of problems you may encounter that we can talk about:

- Having to choose between two ways of handling something
- Having some kind of fight or disagreement with someone
- Having to decide about doing something that might get you into trouble, like taking something or playing with fire

Please write down 1 to 3 problems you wish to discuss next time.

Problems to discuss next time:

1. ______________________________

____

____

2. ______________________________

____

____

3. ______________________________

____

____

Child Form 12.1

Being Assertive

Think of at least one situation that happened at school, at home, or anywhere else where you wish you would have been (or you think you could have been) more assertive, like where you should have asked for something or some help (in a more appropriate way) or stood up for yourself.

Where were you? ______________________________

What happened? ______________________________

What did you do? ______________________________

How did you handle yourself?

1	2	3	4	5
poorly	not so well	okay	well	great

How disappointed or frustrated were you?

1	2	3	4	5
Not at all	A little	Somewhat	A lot	Very much

Was there anything that you did that you wish you hadn't done?

____

Were you being passive or aggressive? ______________________________

What got in the way of being assertive here (that is, asking for something or making a request, or standing up for yourself)?

____

How could you have said something? ______________________________

What would have been the best way to be assertive in this situation?

____

Child Form 12.2

Requests

When making requests, you could use a FAVOR:

F – Find a neutral tone to use (not harsh or loud)

A – Ask, don't demand

V – Verbalize what you want simply and specifically

O – Offer your reason for making this request (and why you are asking them)

R – Remember to be polite (use "please" and "thank you")

Child Form 12.3

Ways People Stand Up For Themselves

Passive **(the "Mouse")**

- Let others boss them around
- People take advantage of them
- Don't stand up for their rights

Aggressive **(the "Monster")**

- Pick on other people
- Take advantage of others
- Don't respect others' feelings
- Only care about getting what they want

Assertive **(the "Me")**

- Let others know how they feel
- Respect other's feelings
- Try to be fair
- Stand up for their rights

Child Form 12.4

"I want"

"I think"

"I feel"

"I" and "You" Statements

"I" Statements:

- Communicate a person's thoughts and feelings
- Involve taking responsibility for your emotions
- Are focused on things you know—how you are feeling
- Focus on what you would like to happen
- Examples: "**I** think this is a bad idea." or "**I** feel frustrated when you interrupt me."

"You" Statements:

- Make the other person feel blamed or attacked
- Lack personal responsibility for emotions
- Are focused on things you don't know—how the other person is feeling
- Focus on what the other person needs to change
- Examples: "**You** have such bad ideas." or "**You** always interrupt me and never listen."

Child Form 12.5

Complaints

If you have a complaint, GET UP!

G – Get the person's attention in a nice way.

E – Explain the situation—what happened.

T – Tell the person what you didn't like—use "I" statements.

U – Under control: do not yell or throw a fit.

P – Pay attention! Listen to the other person's reaction.

About Complaints:

- A positive, assertive complaint can make a situation better.
- An aggressive, angry complaint usually makes a situation worse because the other person doesn't want to change and make it better. Then, everyone ends up mad and frustrated.
- Try to complain only when it really matters.
- Sometimes, when people are really angry, they might forget about giving a complaint in the right way because the other person might not seem to be listening to his or her complaint. But, just because the other person is having trouble listening (or understanding you) doesn't mean that you should give up doing it correctly.

Child Form 12.6

Making Complaints

Think of something that bothered you the past week and that you *should* have complained about using the right way to complain.

Where were you? ______________________________

What happened? ______________________________

What did you do? ______________________________

How did you handle yourself?

1	2	3	4	5
poorly	not so well	okay	well	great

How angry were you?

1	2	3	4	5
burning mad	really angry	moderately angry	mildly angry but still ok	not angry at all

Was there anything that you did that you wish you hadn't done?

____

What got in the way of making an assertive complaint?

____

How could you have said something? ______________________________

What would have been the best way to make a complaint or to be assertive?

____

Appendix C

Parent Forms

Parent Form 5.1

Children and Fire program—Treatment Plan and Goals

Client Name ________________ **Date:** __________ **Number** _______

In collaboration with this program/practitioner, I have agreed to work on the following goals:

____

____

Let's work with your child:

- to understand any involvement with fire, including firesetting incidents
- to learn fire safety education that can be applied in the home
- to identify, express, and manage feelings
- to come up with ways to think about problem situations that can lead to bad decisions
- to solve problems to avoid being involved with fire
- to participate in safe activities (e.g., after-school programs, sports)
- to be assertive rather than aggressive or passive
- other:

Let's work with you:

- to reduce personal/family stress or events that may relate to your child's fire incident
- to learn fire safety education for use at home
- to monitor and supervise your child's whereabouts and activities
- to learn some strategies that help parents feel more confident and effective in managing their child's behavior
- to support safe activities for your child.

- to conduct a home program to manage positive and unsafe behavior
- to develop a plan to discourage your child's contact with fire and increase his or her involvement in more safe and appropriate social activities
- other:

Your strengths / resources that will help you reach your goal are: ______________________________

____

Benefits, risks, and alternatives to treatment discussed with client: ☐ Yes ☐ No

This treatment plan will be reviewed after ____________ sessions, or by ________________ (date).

Client signature: __________________________ Date: ____________

__________________________ Date: ____________

Therapist signature: __________________________ Date: ____________

Supervisor Initials: __________________________

Parent Form 5.2

How to Relax in a Few Minutes Handout

1. Get comfortable in your chair. Close your eyes.
2. Become aware of your breathing.
3. As you breathe in and out, slow your breathing down.
4. As you slow down your breathing, try to count from 1 to 10 as you hold that single breath.
5. Now exhale slowly, counting from 1 to 10 again until you are out of breath.
6. Repeat this deep, slow inhaling and exhaling as you count until you feel relaxed.

Now gradually let your breathing return to normal and open your eyes.

Parent Form 5.3

Behaviors Associated with Children Who Play With or Set Fires and Some Possible Consequences for Managing These Behaviors

Problem Behavior	Possible Consequence
1. Child keeps, hides, or has matches or lighters.	Child loses privilege(s).
2. Child is in possession of other incendiaries (i.e., items capable of catching fire, causing fires, or burning readily).	Consider scheduled or random search of child's clothing or materials to remove all items.
3. Child plays with matches or lighters.	Child is assigned extra chores or work tasks.
4. There are community reports of the child playing with matches or lighters.	Child is assigned extra chores or work tasks to repair any damage he/she caused.
5. Child sets a fire that burns property or people (e.g., toys, household items, other objects, clothes, skin).	Child loses privilege (e.g., going out of house) until the chore or work task is completed.
6. There are community reports of the child burning toys or other objects.	Child makes material (e.g., replaces an item) or financial (e.g., pays for an item) restitution for any damages caused, and/or tries to restore or clean the area (e.g., clean up, fix or replace materials). Teen's behavior is reported to authorities.
	Child loses privilege (s) for increasing lengths of time with every new report.
	If child was with a peer during problem behavior, child briefly loses the privilege of being with that peer (or interactions with peer are supervised).
7. Child does not tell parents his/her whereabouts.	Child loses privilege of having unsupervised time away from his/her house; amount of time is reduced with each additional rule violation.
8. Child is not at the agreed upon location when a parent checks on him/her.	Child's curfew is made earlier.
9. Child is noncompliant when asked by parent to report his/her activities with peers	
10. Child is noncompliant or unable to give details of his/her activities with peers when asked by parent	
11. Child is involved in inappropriate individual/ peer activities	

Parent Form 5.4

Worksheet for Describing Possible Target Behaviors and Consequences to Manage Them

Problem Behavior	Possible Consequence

Parent Form 8.1

Do's and Don'ts for One on One Time

Skills to use during one on one time:

- *Praise*—especially praise for specific behavior.
- *Reflection*—Repeat what your child says to help him or her know you are listening. For example, Child/adolescent: "Now I'm going to put the race car on the track." Parent: "Oh, you're putting the race car on the track."
- *Description*—Describe what your child/adolescent is doing. For example, Parent: "Oh, so now you're building a tower and putting on the highest block."
- *Enthusiasm*—One on one time is always more fun and warm when parents are genuine and enthusiastic!
- *Attention* to positive and neutral behavior.
- *Ignore* minor misbehavior.

Things to avoid during one on one time:

- *Criticism*—Since one on one time is meant to be enjoyable and to promote a positive parent-child relationship, it is preferable to make this activity as rewarding and positive as possible. Try to keep your comments positive, even if you don't agree with some aspect of the activity.
- *Questions*—Questions put pressure on a child/adolescent to respond and can end up directing the interaction. While questions are OK in many other interactions, try to avoid questions during one on one time.
- *Commands*—This is really not the time to direct the child to do things, like chores or other activities. It is a really meant to be a fun, child directed, activity that you do together.
- *Teaching*—Teaching can also direct the interaction, and it changes one on one time from a fun activity into one where the parent is providing education and instruction.
- *Discipline, scolding, or punishment* for minor misbehavior.

Parent Form 8.2

Planned Attending and Ignoring

Rules to Follow When Using Attending and Planned Ignoring

1. Never ignore dangerous behavior when there is likelihood of injury.
2. Always ignore the same behavior consistently. (Don't look at child, speak, smile, or frown when ignoring.)
3. Anticipate that the ignored behavior may increase initially (i.e., get worse).
4. Remember why you are ignoring so you don't feel guilty and then stop ignoring (i.e., it should be planned).
5. Always attend to behavior you want to increase after ignoring a behavior—think about which behaviors to attend to.
6. Be specific if you're praising a desired behavior so the child knows what he or she has done that is positive.
7. Make eye contact when attending but not when ignoring.
8. Smile when attending.
9. Use a pleasant voice when attending.

Parent Form 8.3

Planned Attending and Ignoring Worksheet

List five behaviors that you would like to see your child do more often, and think of ways you could give positive attention to these behaviors.

Desired Behaviors	Ways to Attend to This Behavior
1.	
2.	
3.	
4.	
5.	

Parent Form 8.4

Suggestions for Giving Positive Feedback and Approval to Your Child

"Nonverbal" signs using body language or physical actions:

- Smiling
- Pat on the back
- Putting your arm around your child
- Hug
- Giving or blowing a kiss
- Giving a "thumbs up" sign
- A head nod or wink
- Clapping your hands
- Others?

____

____

Verbal signs you say:

- "I really like it when you . . ."
- "What you just did was wonderful . . ."
- "Great job!"
- "Terrific!"
- "Fabulous!"
- "What a nice thing to do . . ."
- "You did that all by yourself . . . way to go."
- "I am so proud of you for . . ."
- Others?

____

____

Parent Form 8.5

Rewards as Positive Reinforcement

Rewards can motivate your child to learn and use new skills. This handout is designed to help you begin thinking about different kinds of rewards and how to use them successfully. Here are some helpful tips to consider when planning to use rewards as positive reinforcement:

1. It is helpful to ask your child what he or she finds rewarding.
An example "Rewards Menu" has been included at the end of this handout to help you begin brainstorming specific rewards to use with your child. It may be helpful to review this menu, discuss the items together with your child, and then chose a reward.

2. It is important to make sure that the reward is realistic for you to give.
Be certain that you have the means (financial, time, etc.) to provide a reward before you agree to give it.

3. After selecting a reward, make sure your child knows what has to be done to earn it.
Talk with your child about the specific behavior that will earn the reward. Give examples of how and when the child might do the desired behavior and explain how and when the reward will be given. Be sure to clarify if there are any rules regarding the behavior (e.g., your child is expected to clean his or her room with less than two reminders from you, or your child needs to turn off the video game no later than 30 seconds after you have asked).

4. Emphasize that a reward is only earned *after* a specific behavior has occurred.
Once a reward has been decided, it is important that your child understand the "cause and effect" relationship between his or her behavior and the reward. IF your child does the desired behavior, THEN the reward is given. It is important that you and your child both understand that rewards always come *after* the behavior.

5. Rewards should be given consistently!
Once you have told your child that a reward will be given for a specific behavior, you must follow through. Your child's motivation to do the desired behavior will greatly increase when he or she sees that a reward consistently follows it. If you promise a reward and do not provide it, your child may lose interest and/or no longer take you seriously.

6. It is important to vary the reward to keep a child's interest.
A child can become used to, and bored by, the same old reward. A powerful way to vary the reward is to use a Reward Menu. This involves writing down a variety of rewards (we suggest 10–15) on a menu and allowing your child to select one item from the menu when a reward has been earned.

7. **Rewards don't always have to be "extras."**
It is okay to take away privileges your child may take for granted and then allow them to "earn them back." For example, a child may find a video game rewarding. A parent could take this away and make the child stick to a behavior expectation to earn video game time. Earning a privilege that may be taken for granted can be rewarding to a child.

8. **Try to emphasize social and privilege rewards over material rewards.**
Material rewards (e.g., toys, cars, money) often lose their reinforcing value and are very expensive. Many children benefit from more involvement with their parents. Rewards involving time and activities with caregivers can be reinforcing and promote involvement at the same time.

Using a Reward Menu

A reward menu is a list of possible rewards that are likely to be successful in reinforcing your child's positive behavior. So, having input from your child may make it more likely that you'll select items or activities that your child would like to earn. Here is a sample Reward Menu that you can use as a model.

Sample Reward Menu

Younger Child: ____________________________

1. 30 minutes of play time outside
2. 1 extra TV show on a school night
3. Having my favorite meal or snacks
4. Play video games for 30 minutes after dinner
5. Having a friend over for dinner or to play

Older Child/Teen: ________________________

1. Staying out later on the weekend by an hour
2. Having my favorite meal or snacks
3. Going over to my friend's house for an hour
4. Going to the mall on the weekend
5. Attending a sporting activity

Parent Form 8.6

How to Use Effective Instructions

Steps for issuing instructions:

1. Instruct or tell your child to complete the behavior in a manner in which you would like to be asked (nicely). Do not give a choice, do not ask a question! Keep your instructions brief and specific, and issue one at a time.
2. Give the child adequate time to begin, and praise the child immediately when he or she begins and follows your command. ("Good job!", "Thank you for doing as I asked.")
3. If your child does not do as asked, give a reminder and state the consequences for non-compliance.
4. Give your child adequate time to respond and reward his or her efforts or, if your child does not respond, follow through with the stated consequences.
5. Once the consequences have been administered, state your original request again in an appropriate manner.
6. With a younger child, you may need to physically, but gently, guide or lead the child to where he or she needs to be to begin the task.
7. Repeat this process as often as needed.

Adapted from: Walker, C. E., Bonner, B. L., & Kaufman, K. L. (1988). *The physically and sexually abused child: Evaluation and treatment*. New York: Pergamon Press, p. 53.

Parent Form 8.7

Effective Instructions and Approval Examples

Effective Instructions

In the situations below, imagine you are the parent. Put a check by the statement that will be most successful in encouraging your child to do what you want.

1. Bob has a baseball game. You want his toys put away so you say...

_________ Put your toys away, Bob, and we can leave for the game, or

_________ Bob, why don't you put your toys away?!

2. Lou, aged 10, is hitting Terry, aged 6. You say...

_________ I can't stand any more screaming! You're both in trouble if you don't stop fighting.

_________ Lou, stop hitting Terry.

3. Debbie asks you a question in a loud, shrill tone of voice. You say...

_________ Debbie, ask the question again quietly and I will answer.

_________ Do you have to yell at me? Can't you talk like a normal child?

Approving Statements

Which statement would make your child feel approved?

1. Jackie came home from school on time. You say...

_________ I'm really pleased that you made it home on time today!

_________ Well, for once you made it home before dinner.

2. You see John doing his homework. You say...

_________ Nothing. (John should be doing his homework.)

_________ You're working hard tonight, John. You're doing a really nice job.

3. You ask Mary to pick up her toys, which she starts to do. You say...

_________ Thanks, Mary. I like it when I don't have to tell you twice.

_________ You surprise me, Mary. Usually you take forever to do anything.

4. Pat accidentally knocks a cup of juice off the counter. She quickly grabs a towel and begins to wipe up the mess. You say...

_________ Pat, you're so clumsy. Look at the mess you've made.

_________ Good, Pat, clean it up. Next time, watch where you set your juice.

Parent Form 8.8

Using "When-Then" Statements

The purpose of a "When-Then statement" is to encourage or increase the frequency of a desired or targeted behavior that your child doesn't do that often by providing access to a more highly rewarding activity—but only after he or she does the targeted behavior. Here, you tell your child that when he does something first, he then gets to do something he wants next. This was originally called "Grandma's Rule," named for the white-haired lady of folk wisdom who told her grandchildren, "You can have some pie only after you eat your vegetables." Later referred to as the "Premack Principle," this simple contingency is based on the idea that a highly rewarding activity can be used to encourage a less rewarding but important activity. Basically, it involves requiring children to do something good for them—but that they would rather avoid—before they get to do something they really like. For example, you can insist that your child put on a jacket before going outside or that homework be completed before turning on the TV. The key to using a "When-Then" statement is that the privilege or more rewarding activity that the child must wait to do is actually something that the child really wants and is planning to do.

Directions:

1. *Think about the behaviors that you want your child to do even more often—behaviors that you believe are good to do, like cleaning her room or studying a little longer each night. Usually, this is a less rewarding activity to the child. List them in the column labeled "Do this activity first." Next, think of activities that your child looks forward to that can be withheld until the other required activity is done. Usually, this is a highly rewarding activity. List those in the column labeled "Before you get to do this."*

Do this activity first:	***Before you get to do this activity:***
______________________	______________________
______________________	______________________
______________________	______________________

2. The "When-Then" statement works best when you speak to your child in a friendly, positive, and enthusiastic way. Note the following examples:

   Right way:

   - "Let's get those blocks picked up so we can read this story."
   - "Billy, your friends are outside. You can go out just as soon as you straighten up your room."
   - "When you finish your homework, you can then watch the hockey game.

Wrong way:

- "If you don't pick up those blocks, I won't read you this story."
- "Billy, you can't go out until you straighten up your room."

3. If your child tries to go ahead with the highly rewarding activity without doing what you asked, call him or her back and repeat what you said. If the child continues to ignore you, either use time-out or take away a privilege. List privileges you might take away here:

____

____

____

Rewarding activities work best when:

- The activity is highly rewarding to the child: It should help change the behavior.
- It comes after the targeted behavior you want to encourage: It should be earned.
- It is given consistently.

Parent Form 8.9

Possible Positive Behaviors to Encourage and Consequences

Family Name: ______________________________

Therapist: ______________________________

Date: ______________________________

Positive Behaviors	Privileges & Rewards
1. ____________	1. ____________
2. ____________	2. ____________
3. ____________	3. ____________
4. ____________	4. ____________
5. ____________	5. ____________

Bonuses: ______________________________

Special Stipulations: ______________________________

Parents' or Guardians' Responsibilities: ______________________________

Parent Form 11.1

Using Logical Consequences Worksheet

Logical consequences can help correct irresponsible or immature behaviors. When experiencing a logical consequence, your child must deal with an effect that is determined by you, the parent. As you consider a logical consequence, the idea is to "select a punishment that fits the crime." Here are some examples of logical consequences:

Problem	Logical consequence
Child damages a neighbor's property	Child makes restitution by doing chores for that neighbor or replaces the item
Child continues to argue with a parent	Parent ignores the child until quiet or child begins to listen

Directions

1. Identify specific situations or behaviors that can best be handled with logical consequences. List those here: ______________________________

____

____

____

2. Decide what the consequences for these behaviors might be and how they should be applied. List those here: ______________________________

____

____

____

3. You may be concerned about letting your child face logical consequences. What are the advantages and disadvantages of letting your child face those consequences?

Disadvantages	Advantages
______________	______________
______________	______________
______________	______________

Having weighed the advantages and disadvantages, make a list of appropriate thoughts you can say to yourself that will enable you to carry through with the use of logical consequences:

4. Your child will likely test your commitment to upholding the consequences, so be prepared! Often this testing may take the form of pleading for help. Ignore these pleas if the task is appropriate for the child's age. If your child refuses to accept the consequences, use a time-out or remove a privilege—whichever best fits the situation. Describe here how you will handle these situations:

5. Please note that you should not lecture your child nor offer sympathy after the consequence occurs. Treat the issue as closed.

Parent Form 11.2

Removing Privileges Worksheet

Removing privileges can be helpful if your child continues to engage in unwanted behavior after you have given feedback or tried other consequences (e.g., planned ignoring). Loss of a privilege is also an appropriate consequence by itself if your child fails to follow an agreement made between the two of you. Used in this way, it is an appropriate method of discipline for older children and adolescents.

Directions:

1. Decide what behaviors your child does that you would like to deal with by removing a privilege. List these behaviors in the chart below in the first column.
2. The privilege must be something that you can actually deny your child. For example, if your child gets home by 3:00 p.m. but you do not get home until after 5:00 p.m., you probably cannot deny your child the privilege of watching TV in the afternoon. Another consideration is that, ideally, the privilege taken away should not affect others. This means that, if the child is to be denied TV, you should consider what affect this will this have on other family members who wish to watch it. List those privileges your child would miss if the privileges were taken away in the chart below in the second column.
3. The number of privileges lost and the length of time they are removed should be matched to the age of the child and seriousness of the particular behavior. Most parents make the mistake of taking away too many privileges for too long a time. This action turns what might be a good learning experience into one of lingering hostility and resentment. We recommend that a privilege be removed for only 24 hours. In fact, parents often get better results if they use even briefer periods, such as taking away a toy for a few hours or preventing the child from watching a favorite TV show for each episode of swearing or talking back. Indicate the length of time that each privilege will be removed in the chart below in the third column.

Behavior Requiring Privilege Removal	Privilege to Remove	Length of Removal

4. When taking away a privilege, be calm but firm. If your child tries to ignore the restriction, remove an additional privilege. Finally, once the privilege has been lost, the punishment accepted, and the time spent, the privilege should be restored and the incident dropped. Avoid any further lectures or reminders. So, when taking away privileges, remember to: 1) make the loss brief, 2) clarify the specific criterion for loss and return of privilege, and 3) be fair and quick (no extra nagging, etc.).

Parent Form 11.3

Possible Problem Behaviors to Discourage and Consequences

Family Name: ______________________________

Therapist: ______________________________

Date: ______________________________

Problem Behaviors	Negative Consequences
1. ____________________	1. ____________________
2. ____________________	2. ____________________
3. ____________________	3. ____________________
4. ____________________	4. ____________________
5. ____________________	5. ____________________

Parent Form 11.4

Point System Program

Family ____________________ Therapist ________________

Date ______________________

To use a point system, you will give your child a "point" or some form of token (e.g. poker chip, bean) each time the child does the desired behavior. A child who earns a certain number of points or tokens over a period of time (for example, 10 points in 1 day) receives a privilege or other reward. You can also deduct points for identified behaviors you want to discourage. Each target behavior needs to be very specific and clear. And, it is important to review the program with your child in advance.

How to set up a point system:

1. *Decide which behaviors earn points.* Be very specific, so your child knows exactly what to do to earn a point. Also, pick behaviors that can occur often (e.g., daily chores, compliance, playing quietly for 15 minutes) so that your child has the chance to earn points and receive a reward each day. List each behavior in the *Behavior* column under step #2.

2. *Determine how many points can be earned for each desirable behavior. You may choose to:*
   a. Give a different number of points for different tasks (e.g., 5 points for making the bed, 3 points for feeding pets).
   b. Assign points to different parts of each task (e.g., cleaning the bedroom can be broken into three parts—3 points for making bed, 2 points for hanging up clothes, and 2 points for sweeping floor).
   c. Award 1 point each time a particular positive behavior occurs (e.g., 1 point for each time the child complies).

List the number of points that can be earned for each behavior in the *Earned Points* column.

Positive Behavior		***Earned Points***
1. ______________________________	=	__________
2. ______________________________	=	__________
3. ______________________________	=	__________
4. ______________________________	=	__________
5. ______________________________	=	__________

3. *Determine which negative behaviors lose points and how many points can be lost for each undesirable behavior.*

   You may choose to:

   a. Lose a different number of points for different tasks (e.g., 5 points lost for playing with fire, 3 points lost for fighting, 1 point for talking back).
   b. Lose 1 point each time a particular negative behavior occurs (e.g., 1 point for each time the child is noncompliant).

List the behavior in the *Behavior* column and number of points that can be lost for each behavior in the *Lost Points* column.

Behavior		***Lost Points***
1. ______________________________	=	__________
2. ______________________________	=	__________
3. ______________________________	=	__________
4. ______________________________	=	__________
5. ______________________________	=	__________

4. *Select a list of special privileges or other rewards to be earned.* Both you and your child should have some ideas about this. However, common treats and privileges include reading or telling your child a story, playing a game together, getting to pick a special food, receiving a small toy, going to the park, staying up late, watching TV, playing on the computer, letting other children spend the night, or doing a special activity outside the home (e.g., movies, bike riding, a picnic).

   List special privileges and rewards here: ______________________________

   ____

   ____

   ____

5. *Decide how many points must be earned to receive a special privilege or treat, and how often it can be earned.* Be realistic. Set the total number of points to be earned at a level your child can achieve. For example, if your child only minds about half the time now, you might say the child must comply with 6 of every 10 commands. This will help your child learn that behaving appropriately will help get the things he or she likes. Later, once your child's behavior improves, you can gradually increase the number of points needed to earn a reward. Your child should have the opportunity to earn a special privilege at least once a day.

6. *Monitor and award points.* Each time your child does the behaviors listed here, praise him or her and award points or tokens. You should then mark this on a monitoring form. While points are usually indicated by checks or hash marks, some children—particularly young children—prefer to have points represented by happy faces or stickers.

   Explain how you will track points: ____

   ____

   ____

   ____

7. *Review points earned and lost, and either give reward or withhold privileges.* You need to go over the point chart with your child at the end of a prearranged time (or times) each day. If enough points are earned, you will socially reinforce the child and let him or her pick a special treat or privilege from the list. If the child does not have enough points, you will briefly and matter-of-factly tell him or her that no privilege was earned. (Remember: NO SCOLDING!)

8. *If your child quickly "goes broke' (loses more points than earned), review the program with the child and try to determine why he or she isn't earning more points than are lost. Try to adjust the consequences that occur, like the number of points assigned to each positive and negative target behavior.*

____

____

Parent Responsibilities and Consequences: ____

____

Special Stipulations ____

____

For allowance programs:

This system is set up similar to a point system, except that positive behaviors are assigned a monetary value. It is important to consider how difficult the task is, how much the child dislikes completing it, and how important the task is to you when assigning dollar or cent values to each task. Monitor the child's behavior with a tracking sheet and determine how often the sheet will be reviewed and the child paid. Avoid lecturing for uncompleted tasks.

Allowance to be earned: **Earned?** **Yes** **No**

Youth: Parent/Guardian: Provider:

Parent Form 11.5

Potential Problems for Implementing a Home Management Program

Forgetting to follow the plan:
Regardless of your mood or what has just happened that minute or that day, you have to be *consistent* in giving or removing points or privileges. Whatever your child has coming should be done.

Setting up a system that is too extensive:
You don't want to set something up that is too big or hard to manage.

Offering a reward or privilege that is too big:
Really big rewards aren't that helpful because they can't be used that often. Instead, try to use small to moderate ones.

Not having a desirable reward or privilege:
Sometimes it is hard to find an effective reward, one that your child will really work to get.

Poor monitoring of performance:
Monitor the outcomes of your program and be consistent, regardless of the outcome.

Parent Form 13.1

Home Management Program (Contingency Plan)

Family Name: __________ Provider:______________ Date:_______

Positive Behaviors:	Consequence:
1. ____________________	1. ____________________
2. ____________________	2. ____________________
3. ____________________	3. ____________________
4. ____________________	4. ____________________
5. ____________________	5. ____________________

Problem Behaviors:	Consequence:
6. ____________________	6. ____________________
7. ____________________	7. ____________________
8. ____________________	8. ____________________
9. ____________________	9. ____________________
10. ____________________	10. ____________________

Parent Responsibilities:____

Special Stipulations: ____

Youth Parent/Guardian(s) Witness

References

Adler, R., Nunn, R., Northam, E., Lebnan, V., & Ross, R. (1994). Secondary prevention of childhood firesetting. *Journal of the American Academy of Child & Adolescent Psychiatry, 33*, 1194–1202.

Barkley, R. A. (1987). *Defiant children: A clinician's manual for parent training.* New York: The Guilford Press.

Barkley, R. A., & Benton, C. M. (2013). *Your defiant child, second edition: Eight steps to better behavior.* New York: Guilford Press.

Barkely, R. A., & Robin, A. L. (2013). *Your Defiant Teen, Second Edition: 10 Steps to Resolve Conflict and Rebuild Your Relationship.* New York, NY: Guilford Press.

Bloomquist, M. L. (2006). *Skills training for children with behavior problems: A parent and practitioner guidebook.* New York: Guilford Press.

Bumpass, E. R., Brix, R. J., & Preston, D. (1985). A community-based program for juvenile firesetters. *Hospital and Community Psychiatry, 36*, 529–533.

Bumpass, E. R., Fagelman, F. D., & Brix, R. J. (1983). Intervention with children who set fires. *American Journal of Psychotherapy, 37*, 328–345.

Campbell, R. (2014). Playing with fire. *Fire Analysis and Research Department, National Fire Protection Association.* NFPA No. USS17.

Campbell, R. (2014). Intentional Fires. *Fire Analysis and Research Department, National Fire Protection Association.* NFPA No. ATP.

Cole, R. E., Laurenitis, L. R., McCandrews, M. M., Schwarztman, P., & Andrews, R. (1983). *Final Report of the 1983 Fire-Related Youth Program Development Project.* Rochester, NY: New York State Office of Fire Prevention and Control.

Cotterall, A., McPhee, B., & Plecas, D. (April, 1999). *Fireplay report: A survey of school-aged youth in grades 1 to 12.* Surrey, British Columbia: University College of the Fraser Valley.

Cox-Jones, C., Lubetsky, M., Fultz, S. A., & Kolko, D. J. (1990). Inpatient treatment of a young recidivist firesetter. *Journal of the American Academy of Child Psychiatry, 29*, 936–941.

Del Bove, G., Caprara, G. V., Pastorelli, C., & Paciello, M. (2008). Juvenile firesetting in Italy: Relationship to aggression, psychopathology, personality, self-efficacy, and school functioning. *European Child and Adolescent Psychiatry, 17*, 235–244.

Deluty, R. (1979). Children's Action Tendency Scale: A self-report measure of aggressiveness, assertiveness, and submissiveness in children. *Journal of Consulting and Clinical Psychology, 47*(6), 1061–1071.

DeSalvatore, G., & Hornstein, R. (1991). Juvenile firesetting: Assessment and treatment in psychiatric hospitalization and residential placement. *Child & Youth Care Forum, 20*, 103–114.

Faranda, D. M., Katskikas, S. L., & Lim, N. (2001). Communities working together: An evaluation of the intervention program for juvenile firesetters and arsonists in Brown County, Florida. *American Journal of Forensic Psychology, 19*(4), 37–62.

Faranda, D. M., Katskikas, S. L., Lim, N., & Fegley, V. M. (2007). The trial of juvenile firesetting, bedwetting, and cruelty to animals: Establishing prevalence and comorbidity. *American Journal of Forensic Psychology, 25*(1), 51–69.

Federal Bureau of Investigation. (2013). *Uniform crime reports.* Washington, DC: Department of Justice.

Federal Emergency Management Agency. (1979). *Interviewing and counseling juvenile firesetters.* Washington, DC: U.S. Government Printing Office.

Federal Emergency Management Agency. (1983a). *Juvenile firesetter handbook: Dealing with children ages 7 to 13.* Washington, DC: U.S. Government Printing Office.

Federal Emergency Management Agency. (1983b). *Juvenile firesetter handbook: Dealing with children ages 7 to 14.* Washington, DC: U.S. Government Printing Office.

Fleischmann, M. J., Horne, A. M., & Arthur, J. L. (1983). *Troubled families: A treatment program.* Champaign, IL: Research Press.

Gannon, T. A., & Pina, A. (2010). Firesetting: psychopathology, theory and treatment. *Aggression and Violent Behaviour, 15,* 224–238.

Hardesty, V. A., & Gayton, W. F. (2002). The problem of children and fire: An historical perspective. In D. J. Kolko (Ed.), *Handbook on firesetting in children and youth* (pp. 1–13). London: Academic Press.

Henderson, J. L., MacKay, S., & Peterson-Badali, M. (2006). Closing the research-practice gap: Factors affecting adoption and implementation of a chldren's mental health program. *Journal of Clinical Child & Adolescent Psychology, 35*(1), 2–12.

Jacobson, R. R. (1985). The Subclassification of chil firesetters. *Journal of Child Psychology & Psychiatry, 26,* 769–775.

Jones, R. T., Kazdin, A., & Haney, J. I. (1981). Social validation and training of emergency fire safety skills for potential injury prevention and life saving. *Journal of Applied Behavior Analysis, 14,* 249–260.

Jones, R. T., Ollendick, T. H., & Shinske, F. K. (1989). The role of behavioral versus cognitive variables in skill acquisition. *Behavior Therapy, 20,* 293–302.

Kennedy, P. J., Vale, E. L. E., Khan, S. J., & McAnaney, A. (2006). Factors predicting recidivism in child and adolescent fire-setters: A systematic review of the literature. *The Journal of Forensic Psychiatry & Psychology, 17*(1), 151–164. doi: 10.1080/14789940500441501

Kolko, D. J. (1983). Multicomponent parental treatment of firesetting in a developmentally-disabled boy. *Journal of Behavior Therapy and Experimental Psychiatry, 14,* 349–353.

Kolko, D. J. (1985). Juvenile Firesetting: A review and critique. *Clinical Psychology Review, 5,* 345–376.

Kolko, D. J. (1988). Community interventions for childhood firesetters: A comparison of two national programs. *Hospital and Coummity Psychiatry, 39,* 973–979.

Kolko, D. J. (1989). Fire setting and pyromania. In C. Last & M. Hersen (Eds.), *Handbook of child psychiatric diagnosis* (pp. 443–459). New York: Wiley.

Kolko, D. J. (1999). Firesetting in children and youth. In M. Hersen & V. V. Hasselt (Eds.), *Handbook of psychological approaches with violent offenders: Contemporary strategies and issues* (pp. 95–115). New York: Plenum Press.

Kolko, D. J. (2001). Efficacy of cognitive-behavioral treatment and fire safety education for firesetting children: Initial and follow-up outcomes. *Journal of Child Psychology and Psychiatry and Allied Disciplines, 42*(359–369).

Kolko, D. J. (2002a). Child, parent, and family treatment: Cognitive-behavioral interventions. In D. J. Kolko (Ed.), *Handbook on firesetting in children and youth* (pp. 305–336). New York: Kluwer Academic Press.

Kolko, D. J. (2002b). *Clinical assessment of juvenile firesetters and their families: Tools and tips.* New York: Kluwer/Academic Press.

Kolko, D. J. (2002c). Research studies on the problem. In D. J. Kolko (Ed.), *Handbook on firesetting in children and youth* (pp. 33–56). San Diego: Academic Press.

Kolko, D. J. (Ed.). (2002d). *Handbook on firesetting in children and youth.* New York: Kluwer/Academic Press.

Kolko, D. J., & Ammerman, R. T. (1988). Firesetting. In M. Hersen & C. Last (Eds.), *Child behavior therapy casebook* (pp. 243–262). New York: Plenum.

Kolko, D. J., Day, B. T., Bridge, J., & Kazdin, A. (2001). Two-year prediction of children's firesetting in clinically-referred and non-referred samples. *Journal of Child Psychology and Psychiatry and Allied Disciplines, 42,* 371–980.

Kolko, D. J., Dorn, L. D., Bukstein, O. G., Pardini, D., Holden, E. A., & Hart, J. D. (2009). Community vs. clinic-based modular treatment of children with early-onset ODD or CD: A clinical trial with three-year follow-up. *Journal of Abnormal Child Psychology, 37,* 591–609.

Kolko, D. J., & Foster, J. (in press). Firesetting in children and youth. In V. B. V. Hasselt & M. L. Bourke (Eds.), *Handbook of behavioral criminology: Contemporary strategies and issues.* New York: Springer Publishing.

Kolko, D. J., Herschell, A. D., & Scharf, D. M. (2006). Education and treatment for children who set fires: Specificity, moderators, and predictors of recidivism. *Journal of Emotional & Behavioral Disorders, 14*(4), 227–239.

Kolko, D. J., & Kazdin, A. (1986). A conceptualiization of firesetting in children and adolescents. *Journal of Abnormal Child Psychology, 14*(1), 49–61.

Kolko, D. J., & Kazdin, A. (1988a). Parent-child correspondence in identification of firesetting among child psychiatric patients. *Journal of Child Psychology & Psychiatry, 29,* 175–184.

Kolko, D. J., & Kazdin, A. (1988b). Prevalence of firesetting and related behaviors among child psychiatric patients. *Journal of Consulting & Clinical Psychology, 56*(4), 628–630.

Kolko, D. J., & Kazdin, A. (1989a). Assessment of dimensions of childhood firesetting among child psychiatric patients and nonpatients. *Journal of Abnormal Child Psychology, 17,* 157–176.

Kolko, D. J., & Kazdin, A. (1989b). The Children's Firesetting Interview with psychiatrically referred and nonreferred children. *Journal of Abnormal Child Psychology, 17*(6), 609–624.

Kolko, D. J., & Kazdin, A. (1991a). Aggression and psychopathology in matchplaying and firesetting children: A replication and extension. *Journal of Clinical Child Psychology, 20,* 191–201.

Kolko, D. J., & Kazdin, A. (1991b). Motives of childhood firesetters: Firesetting characteristics and psychological correlates. *Journal of Child Psychology and Psychiatry, 32*(3), 535–550.

Kolko, D. J., & Kazdin, A. (1992). The emergence and recurrence of child firesetting: A one-year prospective study. *Journal of Abnormal Child Psychology, 20,* 17–37.

Kolko, D. J., Kazdin, A., & Meyer, E. C. (1985). Aggression and psychopathology in childhood firesetters: Parent and child reports. *Journal of Consulting & Clinical Psychology, 53,* 377–385.

Kolko, D. J., & Kazdin, A. E. (1994). Children's descriptions of their firesetting incidents: Characteristics and relationship to recidivism. *Journal of the American Academy of Child Psychiatry, 33*(1), 114–122.

Kolko, D. J., Scharf, D. M., Herschell, A. D., Wilcox, D. K., Okulitch, J. S., & Pinsoneault, I. (2008). Survey of juvenile firesetter intervention programs in North America. *American Journal of Forensic Psychology, 26*(4), 41–66.

Kolko, D. J., Simonich, H., & Loiterstein, A. (2014). Alternatives for families: A cognitive behavioral therapy. An overview and a case example. In A. Urquiza & S. Timmer (Eds.), *Evidence-based approaches for the treatment of maltreated children* (pp. 187–212). New York: Springer.

Kolko, D. J., & Swenson, C. C. (2002). *Assessing and treating physically abused children and their families: A cognitive behavioral approach.* Thousand Oaks, CA: Sage Publications.

Kolko, D. J., Watson, S., & Faust, J. (1991). Fire safety/prevention skills training to reduce involvement with fire in young psychiatric inpatients: Preliminary findings. *Behavior Therapy, 22*(2), 269–284.

Kuhnley, E. J., Hendren, R. L., & Quinlan, D. M. (1982). Firesetting by children. *Journal of American Academy of Child Psychiatry, 2*, 560–563.

Lambie, I., Ioane, J., Randell, I., & Seymour, F. (2013). Offending behaviours of child and adolescent firesetters over a 10-year follow-up. *Journal of Child Pyschology and Psychiatry, 54*, 1295–1307.

Lewis, N. O. C., and Yarnell, H. (1951). Pathological firesetting (pyromania). *Nervous and Mental Disease Monograph* (Vol. No. 82). Nicolasville, KY: Coolidge Foundation.

Lyons, J. S., McClelland, G., & Jordan, N. (2010). Fire setting behavior in a child welfare system: Prevalence, characteristics and co-occurring needs. *Journal of Child and Family Studies, 19*, 710–727.

MacKay, S., Henderson, J., Del Bove, G., Marton, P., Warling, D., & Root, C. (2006). Fire interest and antisociality as risk factors in the severity and persistence of juvenile firesetting. *Journal of the American Academy of Child & Adolescent Psychiatry, 45*(9), 1077–1084.

MacKay, S., Paglia-Boak, A., Henderson, J., Marton, P., & Adlaf, E. (2009). Epidemiology of firesetting in adolescents: Mental health and substance use correlates. *Journal of Child Psychology and Psychiatry, 50*, 1282–1290.

Mackay, S., Ruttle, E. M., & Ward, A., K. (2012). The developmental aspects of firesetting. In G. L. Dickens, P. A. Sugarman & T. A. Gannon (Eds.), *Firesetting and mental health* (pp. 84–106). London: RCPsych Publications.

Massachusetts Coalition for Juvenile Firesetter Intervention Programs. (2001). Our focus: Kids and fire-intervention and education. *Massachusetts Coalition for Juvenile Firesetter Intervention Programs: Annual Report.* Westport Point, MA: Massachusetts Coalition for Juvenile Firesetter Intervention Programs.

Massachusetts Coalition for Juvenile Firesetter Intervention Programs. (2002). Expanding the circles of care, datalink 2002. *Massachusetts Coalition for Juvenile Firesetter Intervention Programs: Annual Report* (pp. 6–7). Westport Point, MA: Massachusetts Coalition for Juvenile Firesetter Intervention Programs.

Massachusetts Coalition for Juvenile Firesetter Intervention Programs. (2004). Kids in Danger *Media Alert.* Wesport Point, MA: Massachusetts Institute for Juvenile Firesetter Intervention Programs.

Michelson, L., Sugain, D. P., Wood, R. P., & Kazdin, A. E. (1983). *Social skills assessment and training with children: An empirical based handbook.* New York, NY: Springer.

McGrath, P., Marshall, P. T., & Prior, K. (1979). A comprehensive treatment program for a firesetting child. *Journal of Behavior Therapy and Experimental Psychiatry, 10*, 69–72.

McNeil, C., & Hembee-Kigin, T. L. (2011). *Parent Child Interaction Therapy, Edition 2.* New York: NY: Springer.

Morris, E., Wilcox, L., Rubens, S., & Vernberg, E. (October, 2012). *Detecting and understanding youth fire interest and fire involvement in an outpatient population.* Poster presented at the National Conference in Clinical Child and Adolescent Psychology, Lawrence, KS.

Morris, E., Wilcox, L., Schwartz, S., Vernberg, E. (2010, October). *The utility of using an evidence-informed screening protocol to identify juvenile firesetters in an outpatient psychiatric population.* Poster presented at the National Conference in Clinical Child and Adolescent Psychology, Lawrence, KS.

Morris, E., Wilcox, L., & Vernberg, E. (October, 2012). *Using the Children's Firesetting Interview to explore the characteristics of fire-involved youth from an outpatient population.* Poster presented at the National Conference in Clinical Child and Adolescent Psychology, Lawrence, KS.

National Association of State Fire Marshals. (1999). Report on campaign messages, themes and existing intervention programs for juvenile fire-setters. Albany, NY: Author.

National Association of State Fire Marshals. (2001). Juvenile firesetter intervention research project: Final report.

National Fire Protection Association. (1978). *Executive summary report of the Learn Not to Burn Curriculum.* Quincy, MA: National Fire Protection Association.

National Fire Protection Association. (1979). *Learn not to burn curriculum.* Quincy, MA: National Fire Protection Association.

National Fire Protection Association. (1995). Report of the NFPA firesetter practitioners' forum. Baintree, MA.

National Fire Protection Association. (2011). NFPA's Latest estimates of home fires started by playing—2011 (pp. 1). Quincy, MA: National Fire Protection Association.

Okulitch, J. S., & Pinsonneault, I. (2002). The interdisciplinary approach to juvenile firesetting: A dialogue. In D. J. Kolko (Ed.), *Handbook on firesetting in children and youth* (pp. 57–74). San Diego: Academic Press.

Oregon Office of State Fire Marshal. (2002a). Juvenile firesetting in Oregon. Salem, OR: Oregon Office of State Fire Marshal.

Oregon Office of State Fire Marshal. (2002b). *Juvenile with fire: Screening tool.* Salem, OR: Oregon Office of State Fire Marshal.

Pinsonneault, I. (2002). Developmental perspectives on children and fire. In D. J. Kolko (Ed.), *Handbook on firesetting in children and youth.* (pp. 15–31). New York: Kluwer/Academic Press.

Repo, E., Virkkunen, M., Rawlings, R., & Linnoila M. (1997). Criminal and psychiatric histories of Finnish arsonists. Acta Psychiatrica Scandinavia, *95*(4), 318–323.

Rice, M. E., & Harris, G. T. (1991). Firesetters admitted to a maxumum security psychiatric institution: Offenders and offenses. *Journal of Interpersonal Violence, 6,* 461–475.

Schwartz, S., Vernberg, E. M., Morris, E., & Hambrick, E. (August, 2011). *Transporting efficacious interventions to the real world: Clinician input on the training process.* Poster presented at the Annual *Meeting of the American Psychological Association, Washington, DC.

Stewart, M. A., & Culver, K. W. (1982). Children who set fires: The clinical picture and a follow-up. *British Journal of Psychiatry, 140,* 357–363.

Strachan, J. G. (1981). Conspicuous firesetting in children. *British Journal of Psychiatry. 138,* 26–29.

Terjestam, P. Y., & Ryden, O. (1996). *Fire-settings as normal behavior: Frequencies and patterns of change in the behavior of 7–16 year old children.* Research Report (pp. 21–147/96). Karlstad, Sweden: Swedish Rescue Services Agency.

Walker, C. E., Bonner, B. L., & Kaufman, K. L. (1988). *The physically and sexually abused child: Evaluation and treatment.* New York: Pergamon Press.

Wilcox, D. K., & Kolko, D. J. (2002). Assessing recent firesetting behavior and taking a firesetting history. In D. J. Kolko (Ed.), *Handbook on firesetting in children and youth* (pp. 161–176). New York: Kluwer/Academic Press.

Wilcox, D. K., Murphy, L., & Goldstein, J. N. (2000, December 1, 2000). *Object relations and trauma in children and adolescents engaged in firesetting behavior*. Paper presented at the Poster Presentation at The Center for Mental Health Services Research and Training, Cambridge Hospital Department of Psychiatry.

Wilcox, L., Vernberg, E., Morris, E., & Rubens, S. (November, 2012). *Development of a community mental health referral system for youth who misuse fire*. Poster presented at the Annual Convention of the Association for Behavioral and Cognitive Therapies, National Harbor, MD.

About the Authors

David J. Kolko, Ph.D., ABPP, is Professor of Psychiatry, Psychology, Pediatrics, and Clinical and Translational Science, at the University of Pittsburgh School of Medicine. He is Director of the Special Services Unit at Western Psychiatric Institute and Clinic, a program devoted to the development and dissemination of evidence-based practices for children or adolescents served in diverse community settings or systems including juvenile justice, child welfare, pediatric primary care, and mental health. He also directs the SAFETY Program (*Services Aimed at Fire Education and Treatment of Youth*). Dr. Kolko has been involved in the study and treatment of child/juvenile firesetters for nearly 40 years. He was a founding member of the regional and state programs for juvenile firesetters in Pennsylvania. His research publications have described the characteristics, assessment, course, treatment, and follow-up outcomes of firesetting children and their families. He is editor of the *Handbook on Firesetting in Children and Youth* (2002, Academic Press).

As a Licensed Psychologist, he is board certified in Child and Adolescent Psychology by the American Board of Professional Psychology. He is a Fellow of the Society of Clinical Child and Adolescent Psychology (Division 53), the Society for Child and Family Policy and Practice (Division 37), and Trauma Psychology (Division 56) of the American Psychological Association. He served two terms on the Board of Directors of the American Professional Society on the Abuse of Children, was co-chair of its Research Committee, received its Research Career Achievement Award for 2001, and received its William Friedrich Memorial Award in 2014.

Dr. Kolko's treatment research interests include the study and treatment of child physical abuse and family conflict, child behavior disorders, childhood firesetting, and adolescent sexual offending, and the integration of behavioral health services in pediatric primary care and family health centers. His work has been supported by the National Institute of Mental Health, Substance Abuse and Mental Health Services Administration, and National Center of Child Abuse & Neglect. He is the senior co-developer of Alternatives for Families: A Cognitive Behavioral Therapy

(AF-CBT; www.afbt.org), an intervention for family conflict and coercion. In that role, he is involved in the dissemination, evaluation, and refinement of AF-CBT across the United States and internationally.

Eric M. Vernberg, Ph.D., ABPP, is Professor of Psychology and Applied Behavioral Science at the University of Kansas. He has been a core faculty member of the Clinical Child Psychology Program at the University of Kansas since 1993. His research program involves applying and advancing psychological science to solve real world problems faced by children, adolescents, and their families. Dr. Vernberg is the author or co-author of more than 100 journal articles and scholarly book chapters in his area of study. His work has been supported by grants from FEMA/US Department of Homeland Security, the U.S. Department of Education, SAMHSA, and the National Institute of Mental Health. Dr. Vernberg has co-written or co-edited several intervention-oriented books and manuals, including *Helping Children Cope with Disasters and Terrorism* (2002, American Psychological Association), *Psychological First Aid Field Operations Guide* (2006; NCTSN and NCPTSD), *Skills for Psychological Recovery Field Operations Guide* (2010, NCTSN & NCPTSD), and *Preventing and Treating Bullying and Victimization* (2010; Oxford University Press). He is a Licensed Psychologist (State of Kansas) and holds Specialty Board Certification in Clinical Child and Adolescent Psychology from the American Board of Professional Psychology (ABPP).

www.ingramcontent.com/pod-product-compliance
Lightning Source LLC
LaVergne TN
LVHW080041210926
848481LV00011B/3249